Volume 3

Dick Smith

Howard W. Sams & Co.
A Division of Macmillan, Inc.
4300 West 62nd Street, Indianapolis, IN 46268 USA

Parts are available individually from Dick Smith Electronics, Inc., or your local electronics distributor, with the exception of integrated circuit ZN414, which can be ordered from the Dick Smith Electronics Mail Order Center, PO Box 8021, Redwood City, CA 94063

DICK SMITH ELECTRONICS

FUN WAY INTO ELECTRONICS

Join the Thousands of People World-Wide who have discovered the Exciting World of Electronics through these entertaining, educational projects!

FUN WAY 1

VOLUME 1

The ideal introduction for beginners from 6-66!

Many people regard electronics as a difficult, mysterious subject, but the Fun Way series makes electronics easy to understand - in short, fun! Volume 1 assumes you know absolutely nothing and gives you the basics: learn what all the components do, what they look like, and how to connect them into a circuit. No soldering is required, so the projects are safe for even the youngest enthusiast!

Book only **$9.95** Cat B-2600

Enables you to build any of the first ten projects in Fun Way One. And because the **components** are not soldered, they are all re-usable so you can build any of the first ten projects too. **$7.25** Cat K-2600

Contains the more specialised components required to complete the last ten projects (11-20) in Fun Way One. **Note:** you will also need the 1-10 kit to build the projects. **$7.90** Cat K-2610

Continuity Indicator	Sound Effects Generator
Transistor Tester	A Crystal Set
Water Indicator	One Transistor Amplifier
Light/Dark Indicator	Beer Powered Radio
The Flasher	Two Transistor Amplifier
An electronic Siren	World's Simplest Transmitter
Dog and cat communicator	Voice Transmitter
A decision maker	CB Radio Receiver
Morse code communicator	Amateur Radio Receiver
Music Maker	Radio Booster Amplifier

Fun Way 1 Gift Box

Everything you need

only **$18.95** Cat K-2605

Fun Way 1 & all the components in one package - everything needed to build any of the Fun Way 1 projects. Great Gift!

FUN WAY 2

VOLUME 2

You've built all the projects in Fun Way 1, and now you're ready for something even more exciting: Fun Way 2! 20 new projects that take up where Fun Way 1 left off: learn how to solder onto a professional printed circuit board, how to use a multimeter, and more! These projects have been chosen especially for their usefulness. check 'em out!

The kits at right do NOT contain instructions. Fun Way Volume 2 is required.

Book only **$9.95** Cat B-2605

Fun Way 2 Gift Pack

$17.50 Cat K-2620

Fun Way 2 book, a quality DSE soldering iron, solder, a 9V battery & our most popular kit, the wireless microphone a $23 value purchased separately yours for less than $20!

Flasher
Make it as the latest in electronic jewellery, or a burglar warning light, etc. The choice is yours and it's easy! Cat K-2621 **$2.95**

Ding Dong Doorbell
Yes, it actually sounds like a great old fashioned doorbell. But it's electronic: and you can build it! Cat K-2622 **$4.50**

Morse Code Trainer
You can turn this audible alarm into just about anything requiring sound. Even for Morsecode – and it's fun to learn! Cat K-2623 **$4.50**

Universal Timer
Eggs too hard? Time them with this great little timer. It really works – set it for a few seconds to 15 mins. Cat K-2624 **$5.50**

Electronic Dice
Just as much fun using it as it is building it! And it is really random – no cheating with this dice! Cat K-2625 **$4.95**

Mono Organ
Snazzy little circuit uses the back of the PCB as the keyboard! Play a tune – it's easy. Cat K-2626 **$7.95**

Pocket Transistor Radio
Build this set and learn about the fundamentals of radio. And then you'll have fun using it too! Cat K-2627 **$7.95**

Touch Switch
Ever wondered how lift buttons work? Build this kit yourself and you'll be able to work it out. Cat K-2628 **$5.50**

Mosquito Repeller
Do mozzie repellers really work? Build this one and find out. You won't hear it – but the mozzies will! Cat K-2629 **$4.95**

Intercom Amplifier
Use it to make the intercom (project 18) or use as a general purpose amplifier. Really handy, good sound output too. Cat K-2630 **$6.50**

Wireless Mic.
Just like the 'bugs' used in spy movies! Transmits to any standard FM radio in another room or next door, etc. Hear it all! Cat K-2631 **$6.50**

Light Activated Switch
Useful in many security applications around the home etc. You could even use it to trigger a burglar alarm. Cat K-2632 **$5.50**

Metal/Pipe Locator
Extremely handy for locating pipes, wires, etc. in walls before drilling holes. Surprisingly simple for such an effective circuit. Cat K-2633 **$6.50**

Sound Activated Switch
Picks up sound and triggers. Use as a telephone bell extender or in security applications. Cat K-2634 **$6.95**

Home/Car Alarm
Very practical! You could use this alarm to protect your home and property! Uses any of the normal alarm triggering devices. Cat K-2635 **$6.50**

Electronic Siren
Just what you need with the alarm! Also makes a great sound effects circuits – and Fun Way 2 tells you what to do! Cat K-2636 **$4.95**

LED Level Display
Hook it up to your stereo and it can warn you of dangerous overloads. Looks pretty nifty. The LED's light up in time to the music! Cat K-2637 **$8.95**

Intercom Unit
Build you own intercom & use it around the home. Practical, simple to build with the intercom amplifier (Project 10). Cat K-2638 **$8.95**

Led Counter Module
An introduction to the world of IC's. This useful gadget can be used for counting applications – or just for fun! Cat K-2639 **$7.95**

Short Wave Receiver
Tune in to short wave stations or emergency services and amateurs etc. Easy and fun to build. Cat K-2640 **$6.95**

SPECIAL NOTICE: These kits do not include instructions. The projects published in the Fun Way books contain entire circuit panels & far too much information to be reprinted for inclusion in each kit. Since nearly all kit customers have already purchased the books, duplicating this information would only increase the price of the kits substantially.

OVER HALF A MILLION FUN WAY KITS SOLD WORLD-WIDE!

When you've completed the Fun Way into Electronics course, you'll probably be eager to go on to bigger & better electronics projects. Why not send for the giant DICK SMITH ELECTRONICS Catalog? It's 148 colorful pages packed with more great kits, components, tools, equipment - everything for the electronics enthusiast! Just send us $1 for postage, or check the box in the order form and we'll send one free with your order!

18 YEARS OF RELIABLE MAIL ORDER SERVICE

DSXPRESS

Name ______________________ Phone ( ) ____________

Organization (if applicable) ______________ Date __________

Address __________________________________

City _________________ State _______ Zip _______

PAYMENT: ☐ CHECK ☐ MONEY ORDER ☐ C.O.D.($1.90) ☐ VISA ☐ MC **Send me** ☐ **a catalog**

CARD # ☐☐☐☐☐☐☐☐☐☐☐☐☐☐☐☐

SIGNATURE ________________ EXPIRES _______

SHIPPING INFORMATION

ORDER ON REVERSE SIDE

SEND TO: DICK SMITH ELECTRONICS P.O. Box 8021 Redwood City, CA 94063

FUN WAY 3 VOLUME 3

By the time you've finished volumes 1 & 2, you're an advanced hobbyist - you're ready for Fun Way 3! Now you'll learn about integrated circuits in 10 fascinating new projects. These projects are more complex than the previous ones (and more interesting!) so there are more detailed explanations, more to build, more fun!

Book only
$9⁹⁵
Cat B-2610

Fun Way 3 Bonus Pack

Get the Electronic Cricket Kit & the Mini Amp Kit (two of our most popular projects) and we'll include the Fun Way 3 book for FREE! This is an excellent gift, ideal for school projects!

only $24⁹⁵
Cat K-2670

Light and Sound
Another multi-talented project. Sound effects, flashing lights - continuity tester, Morse code oscillator . . . it's got the lot!
Cat K-2665 **$8⁹⁵**

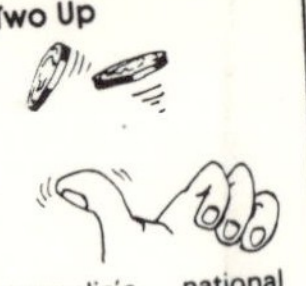
Two Up
Australia's national game - now done electronically because the original King George pennies are hard to get!
Cat K-2661 **$9⁹⁵**

Cricket
Whaaaa! Hide it in a dark room and it starts chirping. Turn the light on and it stops. It's infuriating!
Cat K-2663 **$9⁹⁵**

Mini Stereo Amp
Just the shot for 'Walkie' stereo - now you can listen to it in your bedroom through speakers! Or make yourself a PA amplifier.
Cat K-2667 **$14⁹⁵**

Minder
A multipurpose project for the car. 'Lights on' warning, 'door open' warning plus a pseudo burglar alarm.
Cat K-2660 **$7⁹⁵**

Mini Colour Organ
Like a disco - but battery operated so it's safe! Connect to your radio, etc. for a LED lightshow.
Cat K-2664 **$12⁹⁵**

Combination Time Lock Switch
Make it into a game . . . or use it as a real alarm component. It's that good! Just try getting into this one!
Cat K-2666 **$12⁹⁵**

Lil Pokey
It's a barrel of fun to play. And you don't risk losing your shirt like the real thing.
Cat K-2662 **$19⁹⁵**

Binary Bingo
A game of skill and reaction time - and a valuable teaching aid! It might sound pretty simple to play - but try it and see!
Cat K-2668 **$19⁹⁵**

Mini Synth
Wow! A real musical synthesiser, with decay, frequency doubling & halving, tremolo . . . unique circuit uses YOU as the note generator!
Cat K-2669 **$19⁹⁵**

The Kits above do NOT contain instructions. Fun Way Volume 3 is required.

SCHOOLS!

Join the many educators around the world using the Fun Way Into Electronics books. Developed by Dick Smith Electronics in response to a tremendous demand from parents & educators for an effective introduction to electronics. Fun Way Into Electronics is based on the concept that learning (as well as electronics should be FUN!

These books lead students through the basics of electronics with practical projects that actually work. The projects are quite safe (using a 9V battery or no battery at all!) The first 20 projects in the series use a unique breadboard that allows beginners to connect components without soldering.

Write us on your school letterhead, and we'll send you information on SCHOOL QUANTITY DISCOUNTS.

Can't decide which Fun Way Kits you want?
Fun Way JUMBO Gift Box

Get all 3 Fun Way books, a soldering iron and a selection of 30 projects - nearly a hundred dollar value if purchased separately - in one great gift box! Hours, days, weeks of fun & learning!

$69⁹⁵
Cat K-2690

ORDER TOLL FREE!
1-800-332-5373

California Residents please call 1-415-368-1066
MONDAY - FRIDAY 10am-6pm SUNDAYS 12am-5pm Pacific Time

Our order lines are open 7 days a week, so you can always reach us to place an order. Use VISA or MasterCard, or pay C.O.D. - your order will usually be on its way within 24 hours. Or use the handy order form below!

For inquiries, call 1-415-368-1071 between 7:30 am & 4:00pm Monday-Friday

FUN WAY ORDER FORM

DESCRIPTION	CAT #	PRICE	QTY	TOTAL	DESCRIPTION	CAT#	PRICE	QTY	TOTAL
1. Fun Way Vol 1	B-2600	$9.95			21. Elec. Siren	K-2636	$4.95		
2. Fun Way 1-10	K-2600	$7.25			22. LED Display	K-2637	$8.95		
3. Fun Way 11-20	K-2610	$7.90			23. Intercom	K-2638	$8.95		
4. Fun Way 1 Gift Box	K-2605	$18.95			24. Counter Module	K-2639	$7.95		
5. Fun Way Vol 2	B-2605	$9.95			25. Receiver	K-2640	$6.95		
6. Flasher	K-2621	$2.95			26. Fun Way 2 Gft Pck	K-2620	$17.50		
7. Doorbell	K-2622	$4.50			27. Fun Way Vol 3	B-2610	$9.95		
8. Morse Code	K-2623	$4.50			28. Light & Sound	K-2665	$8.95		
9. Timer	K-2624	$5.50			29. Two Up	K-2661	$9.95		
10. Elec. Dice	K-2625	$4.95			30. Cricket	K-2663	$9.95		
11. Mono Organ	K-2626	$7.95			31. Stereo Amp	K-2667	$14.95		
12. Pocket Radio	K-2627	$7.95			32. Minder	K-2660	$7.95		
13. Touch Switch	K-2628	$5.50			33. Color Organ	K-2664	$12.95		
14. Repeller	K-2629	$4.95			34. Time/Lock Switch	K-2666	$12.95		
15. Intercom Amp	K-2630	$6.50			35. Lil Pokey	K-2662	$19.95		
16. Wireless Mic	K-2631	$6.50			36. Bingo	K-2668	$19.95		
17. Light Switch	K-2632	$5.50			37. Mini Synth	K-2669	$19.95		
18. Pipe Locator	K-2633	$6.50			38. Fn Wy Bonus Pack	K-2670	$24.95		
19. Sound Switch	K-2634	$6.95			39. Fun Way JUMBO	K-2690	$69.95		
20. Home/Car Alarm	K-2635	$6.50					TOTAL		
							CA SALES TAX		
							HANDLING		
							SHIPPING		
							COD		
							AMOUNT DUE		

Be sure to fill out the address info on reverse side.
MINIMUM ORDER $10

You can buy your kits at

Retail Centers

390 Convention Way
Redwood City, CA 94063
(415) 368-8844

2474 Shattuck Avenue
Berkeley, CA 94074
(415) 486-0755

1830 Westwood Blvd.
Westwood, CA 90025
(213) 474-0626

4980 Stevens Creek Blvd
San Jose, CA 95129
(408) 241-2266

Watch for more stores OPENING SOON!

Contents

About the Fun Way series

You may have already been introduced to the exciting world of electronics through our other publications, Volumes 1 and 2. Fun Way 1 is the first in the series and if you are a complete beginner, we recommend that you start with this and try a few of the experiments. This gives some insight into the use of electronic components, and is based on a re-usable breadboard approach. This allows you to perform many experiments without destroying components. No soldering is involved so specialised skills are not required to finish a project. All instruction material is very graphic so that circuit reading or knowledge of electronic principles is not necessary.

From this basic beginning, you advance to Fun Way 2. The greatest step made here is that all the projects are constructed on conventional printed circuit boards, the foundation building block of modern electronics. Each project has a specific board and all components are soldered into circuit. This is where the first real skill is required, that of soldering. A little practice is required but within a short period of time, any original difficulties should be overcome and real progress made.

Circuit diagrams and component use is explained. How project electronics work, wiring and variation of circuit detail are given, yet all material is kept simple. Circuit board overlay and wiring detail is very graphic, nothing is taken for granted. Even without knowledge of electronic theory, all projects can be built and operated by the beginner.

Fun Way 3 is the next natural progression in the series. The physical detail of printed circuit board layout is less graphic and follows more conventional drafting techniques that are more likely to be encountered in general electronic practice. The finer detail of wiring and interfacing is in some cases left to the constructor. Some of the projects can be put to various uses so some basic knowledge would have to be assumed.

This series has a higher educational level yet all projects are still exciting. The reward level of a complete, working model is high. Many hours of enjoyment and fun can be guaranteed with these projects, both during construction and in use.

The projects range in complexity from very simple to moderate. Some are games of chance, some novel, while others have a more dedicated use. Where required, we have included housing detail and have included front dress panels in the rear of this book to give your project a professional finish. Each project is based around an Integrated Circuit (IC), either digital or linear, and these range from simple timers to operational amplifiers to more complex digital counters.

Good reading and construction . . .

Welcome to my

Fun Way Into Electronics Volume Three

With this latest in the Fun Way series, you'll discover just how far electronics has come in the past few years. Because this book is devoted entirely to using the very latest in electronics – integrated circuits.

Who would have imagined when I started building electronic projects about thirty years ago that one day just about every component in the project would be available in a single tiny package.

Just imagine the advances over the next few decades!

In fact, it was modern electronics which enabled me to fly my helicopter solo around the world.

From the radio transceiver which kept me in touch with amateur operators all around the world, through to my Omega navigation system which pin-pointed my position with incredible accuracy: all made possible only by modern electronics.

You'll have the opportunity to build some really exciting projects in Fun Way 3, and you'll be learning about the devices and the way they work while you're building them. In today's world (and tomorrow's!) an understanding of technology is very important.

And like the name of the book says – it really is electronics the Fun Way.

Dick Smith

VK2DIK

Tools

Practical electronics is no different from any other manual skill. To make the job easier and to achieve a professional finish, you require the correct tools. Although there is no hard and fast rule with the type of tools you require for printed circuit board and project assembly, it pays to buy quality. With the projects in this series, the following tools will help you build and finish the jobs with a minimum of fuss. The little extra money spent on quality tools will be well rewarded with the personal satisfaction of a job well done, and the tools will last a lot longer.

1. PRINTED CIRCUIT BOARD ASSEMBLY

Wire Cutters

A precise, sharp, flush side cutting nipper with a small work head is required for component pigtail and wire cutting with normal PC board assembly. The Dick Smith T-3205 is ideal for this purpose.

Long Nose Pliers

A fine pointed long nose plier without serrations on the jaw is required to handle legs of components during the assembly of boards. The slender needle nose types are more applicable for this type of work where close, precise bends and manipulation is required. A good example of the type required is the Dick Smith T-3570.

2. WIRE STRIPPING AND CUTTING

Where hookup wire is to be cut and stripped, a heavier duty side cutting plier is needed. The Dick Smith T-3220 suits this purpose. It combines a sharp positive cutting edge and a wire stripping facility.

A convenient wire stripping tool may be sought where hookup wire of varying sizes has to be stripped. The Dick Smith T-3630 adjustable stripper caters for all common wire types.

3. MECHANICAL HANDLING

Pliers

Where it may be necessary to bend metal or shape mechanical parts, a heavier duty type of plier may be required. The Dick Smith T-3325 flat nose and T-3565 would be useful in this area.

Screwdrivers

Both conventional blade and Phillips screwdrivers are required for general work. The Dick Smith T-4090 150mm blade driver and the T-4040 Phillips type are well representative of this requirement. For smaller work on PC boards, pre-set pots for example, the T-4360 6 piece precision set is a convenient pack.

Wire Cutters Cat T-3205

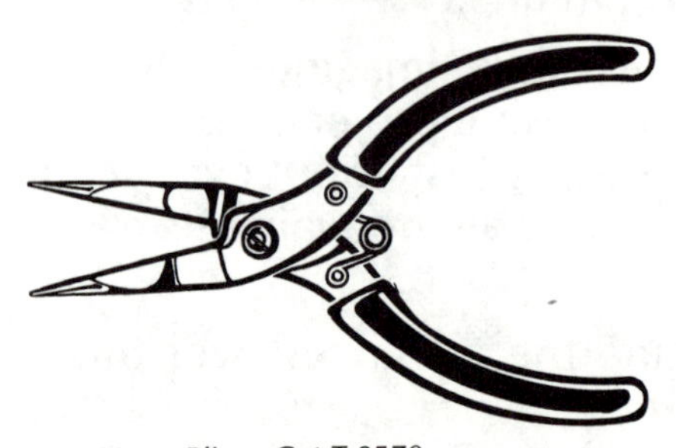

Long Nose Pliers Cat T-3570

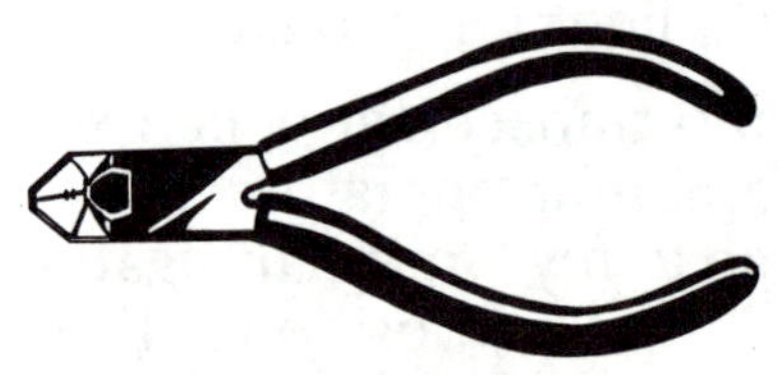

Wire Strippers Cat T-3220

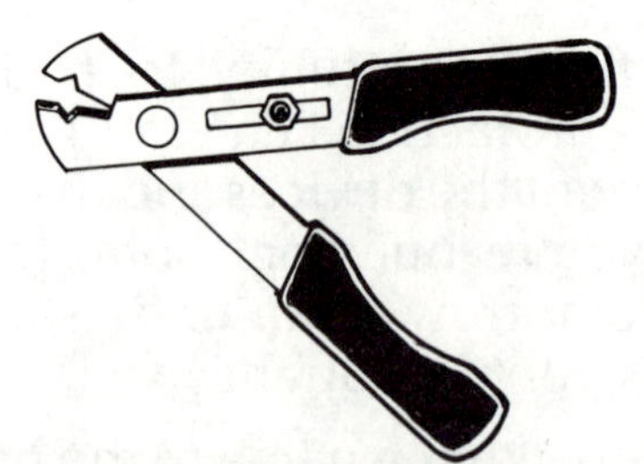

Adjustable Wire Strippers Cat T-3630

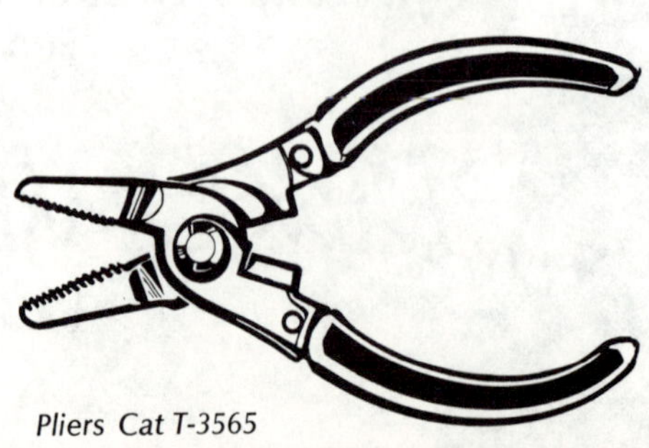

Pliers Cat T-3565

Medium Screwdriver Cat. T-4090

Phillips Head Screwdriver
Cat. T-4040

Drills – P.C Boards

Where constructors fabricate their own PC boards, small drills are required to make component holes. Two convenient sizes to have in your stock are the T-4820 1mm and T-4825 0.8mm. For general work, a set of drills from 1.5mm to 6.5mm are a handy range. Other sizes that are commonly used are 7mm and 9mm.

Power Unit

To turn these drills, you require a high speed power unit. The Dick Smith T-4751 suits this purpose.

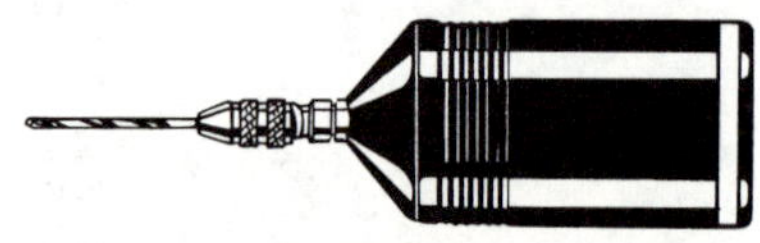

Power Unit Cat T-4751

Files

For shaping and final finishing of panel work, a set of needle files is a worthwhile addition to the hobbyist toolbox. The Dick Smith T-4960 set is a good example.

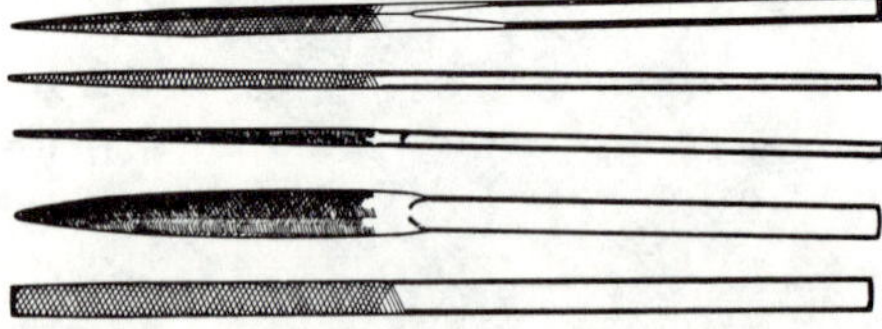

File Cat T-4960

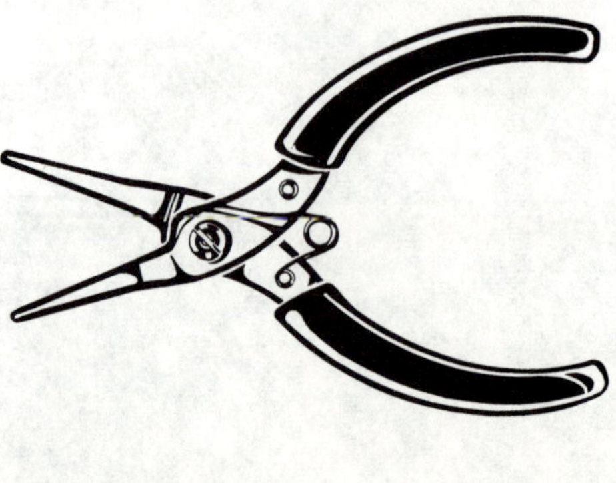

Flat Nose Pliers Cat T-3325

Saws

A piercing or jewellers saw is another handy addition to the toolbox. This enables small precise holes to be cut in thin metal such as that found in project boxes and cases. They are ideal for cutting front panels of Zippy Boxes.

Knives

There are very few projects that can be finished without the need for some form of sharp bladed knife. For trimming and finishing, these are indispensable. A replacement blade type is the more versatile. A popular adjustable blade trimming knife is the Dick Smith T-5220.

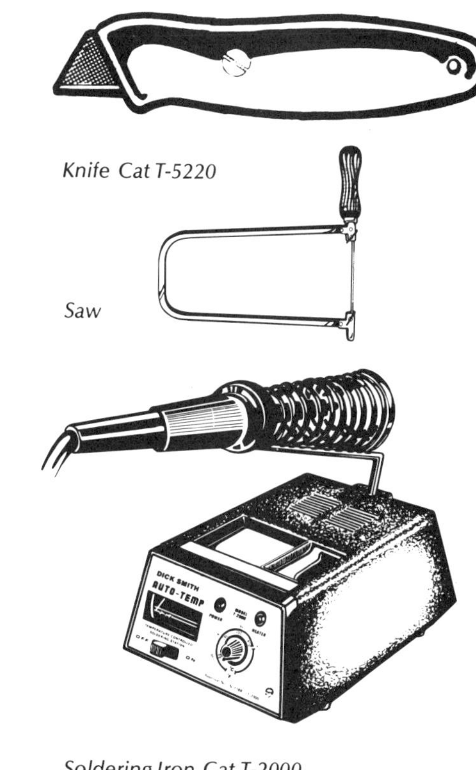

Knife Cat T-5220

Saw

Soldering Iron Cat T-2000

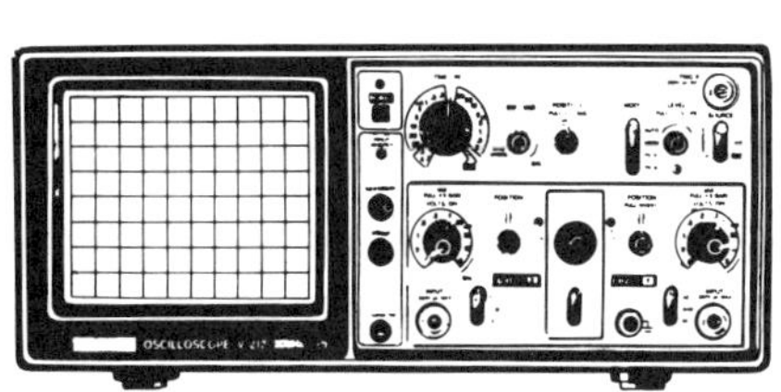

Multi-purpose Stand Cat T-5700

General Purpose Iron Cat T-1300

Analogue Multimeter Cat Q-1140

Oscilloscope Cat Q-1240

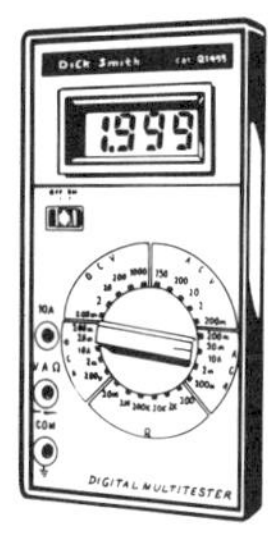

Digital Multimeter Cat Q-1500

4. SOLDERING IRONS

For general purpose PC board assembly and associated wiring, a smaller power and size soldering iron is required. Units in the 10 to 16 Watt class are ideal. An iron of 30 Watts would be considered the maximum for an inexperienced user. A temperature controlled iron system is now looked upon as a common requirement of professionals and hobbyists alike for the assembly of PC boards. For general work, we suggest the Dick Smith T-1330, while for printed circuit boards, the T-1820 Adcola 12 Watt and the T-2000 Dick Smith temperature controlled system.

Iron Tip

The size of the tip is somewhat dependent on the area of the joint or pad to be soldered. Generally, a conical or chisel shaped tip with a surface no greater than 4mm across will suffice. Obviously, a pad size of 2.5mm diameter closely positioned next to tracks would make a 4mm tip hard to use. A working surface around 2mm to 2.5mm would be more applicable in this case.

Soldering Aids

Where a PC board is loaded with components ready to solder, some means of holding it during the soldering operation may be desirable. The Dick Smith T-5700 multi-purpose stand makes a handy aid. It not only holds the PCB, but also your iron and solder roll. also your iron and solder roll.

5. TEST INSTRUMENTS

For any hobbyist seriously interested in electronics construction, a multimeter is almost a necessity. When working with semiconductors, a basic sensitivity of not less than 20k ohms per volt is preferred for accurate readings. A sensitivity of 100k ohms per volt is even better and the Dick Smith Q-1140 analogue multimeter with inbuilt transistor tester is ideal. For greater accuracy and decisive readings, a digital multimeter may be considered. The Q-1455 or Q-1460 units are good examples of instruments of this type.

Although it may be out of the financial reach of some experimenters, an oscilloscope can add another world to the serious hobbyist. Its versatility is probably not realised until one is used in a practical situation. The Dick Smith Q-1243 represents good value for such a versatile instrument.

Components you will use in these projects

The symbols we have shown next to the components in the component descriptions are standard symbols used, with minor variations, throughout Australia. Some countries, particularly European, use different symbols, but it doesn't take too long before you can work out what any symbol means. First of all, become fully conversant with the symbols we use in Australia, and you should have no troubles.

One problem which beginners – and even experts – often have when looking at circuit diagrams is that some people use different methods of marking connections and cross-overs on circuits. Two circuits, side by side, may use the same thing to mean opposites! We have used the standard system showing joined lines with a dot.

If you find a circuit which uses a different system, just remember that it will be a standard throughout the circuit diagram. So look for an area which you know must have joins and cross-overs and, with that information, you'll be able to work out the circuit in no time at all!

COMPONENT	WHAT IT LOOKS LIKE	CIRCUIT SYMBOL	WHAT IT DOES

BATTERIES & BATTERY CLIP

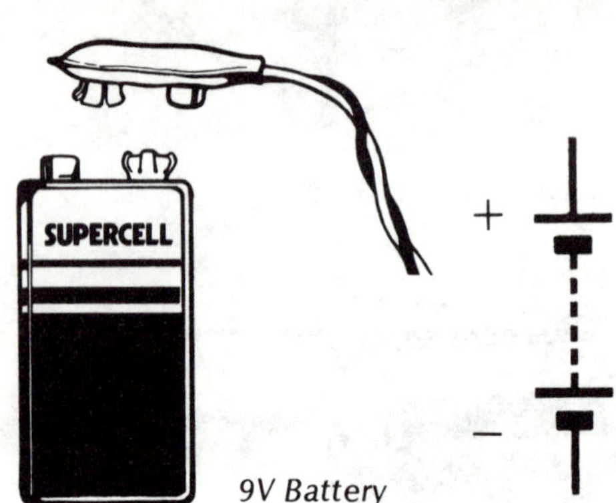

9V Battery

1.5V Dry Cells

The batteries we use in the projects described in this book are DRY CELLS – i.e. they produce an electromotive force or voltage through chemical reaction between a relatively dry paste of chemicals, the zinc case of the battery and a carbon rod. Thus they may be turned upside down without fear of leakage. However, they may NOT be recharged in the way ACCUMULATORS can be (the type you find in motor vehicles).

All batteries are polarised, with a positive and a negative terminal, and must not be connected the wrong way around in a circuit.

Some of the circuits have been designed to use the readily available type 216 9 volt transistor radio battery (Dick Smith Cat S-3110 or similar). These should be used with a battery clip (Cat P-6216) to prevent incorrect connection of the battery. This clip only fits on one way: the red lead is the positive, and the black lead is the negative.

CAPACITORS:

Capacitors store electric charges. The higher the capacity, the more electric charge the capacitor can store. Capacitance is measured in microfarads (uF) and picofarads (pF).

Capacitors are marked with both their capacitance value and a voltage rating. If this voltage rating is exceeded, the capacitor can be seriously damaged. However, it is almost always permissable to use a capacitor with a **higher** voltage rating than the one called for. For example, if a circuit specifies a 1uF 10 volt capacitor, you could use a 1uF 15 volt, 50 volt or even 1,000 volt (if you could find one!) without any problems. You probably could not, however, use a 1uF 9 volt or anything lower.

POLYESTER CAPACITORS:

Often called 'greencaps' because they are usually green, these capacitors are used mainly in audio circuits. They range in value from .001uF to around a few microfarads. They are not polarised.

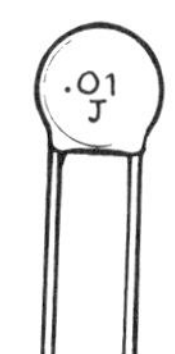

DISC CERAMIC CAPACITORS:

These look like a small disc – hence the name. They range in value from 0.47 uF or so down to 1 pF. You can often use a disc ceramic when a polyester is called for, but the reverse is not always the case. They are not polarised.

ELECTROLYTIC CAPACITORS:

"Electrolytics" for short!

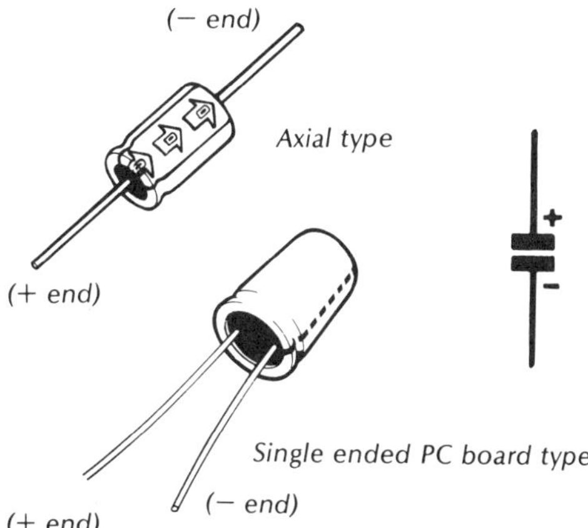

Electrolytics are polarised and they are normally marked so that you cannot mix up the connections. The negative terminal is usually the one marked to indicate polarity. They range in value from around 0.5 uF up to hundreds of thousands of microfarads. They have two leads, which may both come from the same end, or one from each end. The type that has two leads protruding from one end is especially designed to be used on printed circuit boards, taking up less space. Both types have the same symbol – a capacitor symbol with a plus and minus indicating polarity.

TANTALUM CAPACITORS:

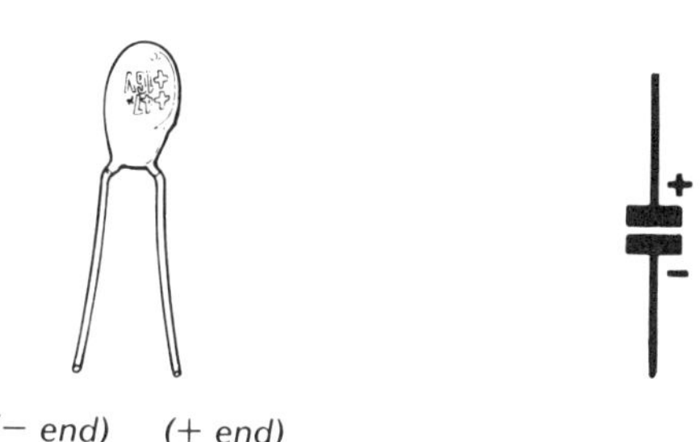

These capacitors offer the great advantage of a high capacity in a very small pack and their radial leads make them ideal for use on PC Boards. Tantalum capacitors are commonly available in values from 1 uF to 100 uF and are usually polarised. They may be used in place of electrolytic capacitors of the same values. Like electrolytics the are polarised.

VARIABLE CAPACITORS:

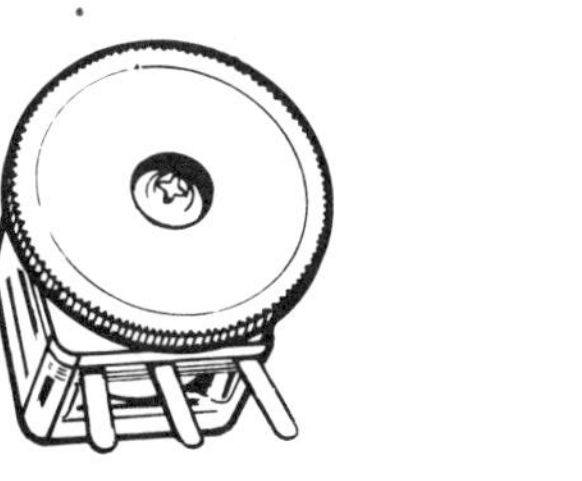

As the general name suggests, the value of these capacitors can be varied. Depending on the type, the change may be effected by a shaft, knob or simple screwdriver adjustment. The well known and a common representative of this group is the tuning component in radio receivers and transmitters. Modern, smaller transistor receivers use solid dielectric tuning gangs (the common name given to these capacitors) while domestic type receivers and transmitters use the air dielectric type. Smaller types of variable capacitors are usually screwdriver adjusted and are pre-set. These types are called trimmers, the capacitance value being normally less than the tuning gang type.

RESISTORS:

Are neither insulators nor good conductors: they are somewhere in between, allowing some current to flow. The lower the resistance, the more current can flow. Resistance is measured in ohms. Resistors are used to limit current to values which can be used by the various components; too much current and the components may be damaged.

Resistors also have a 'power' rating; they must not be called upon to pass too much current or this power rating is exceeded and the resistor may be damaged or destroyed. Our projects use resistors between ¼ and 1 watt power rating. However, like capacitor voltage ratings, it is in order to use higher rated components.

FIXED VALUE RESISTORS:
Various values, ¼ – 1 watt

Fixed value resistors are marked with a colour code, given on page 21 so you will be able to identify each resistor used in these projects.
Resistors are not polarised.

VARIABLE RESISTORS:

or 'potentiometers' (pots)

A variable resistor, or potentiometer to give it the correct name (often referred to as a pot), is merely a fixed value resistor with a slider arm (or 'wiper') that allows continuous adjustment of the value with respect to either end. The action varies with different types. The adjustment of the wiper may be by a hand operated shaft or slot for screwdriver. The shaft operated type is known to all as the volume and tone controls on radios and stereos. Screwdriver adjustment types are known as trimmers and are associated with

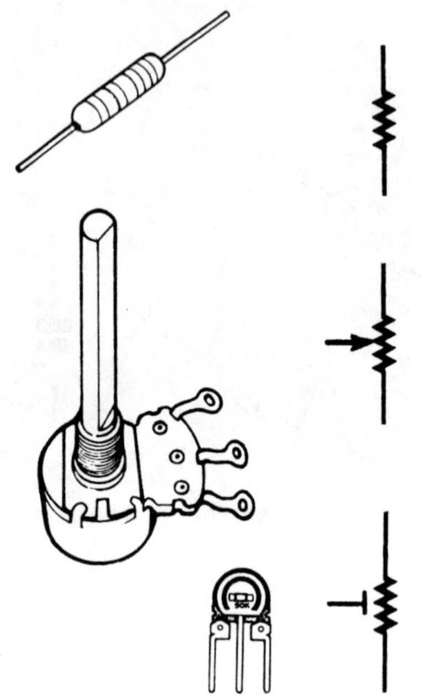

circuits that require variations in parameters only on odd occassions. PC board mounting types are typical of this form.

The wiring of potentiometers with respect to the wiper has to be considered. With a volume control for example, the connections to the circuit have to be made so that the volume is increased as the shaft is turned clockwise. These connections are sometimes marked, W for wiper, C for the clockwise and CW for counter clockwise limits.

LIGHT DEPENDENT RESISTOR (LDR):

(Dick Smith Cat Z-4802)

The active base of this component is formed on the flat of a body so that it can be exposed to light. The chemical structure of this 'resistive' material is different from that of the

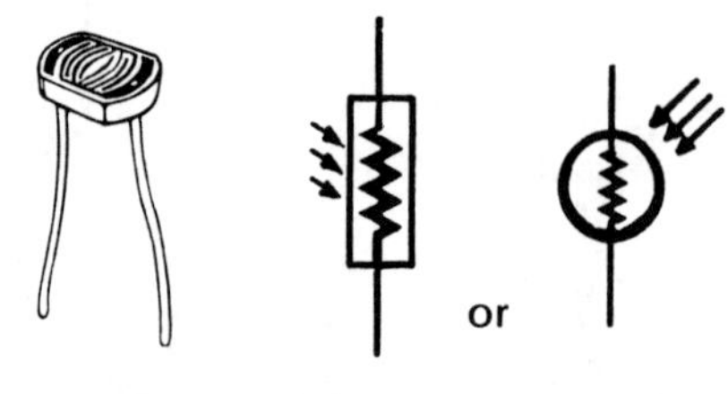

carbon or metal film type in that it effectively lowers in resistance as the light level on this surface increases. Cadmium sulphide is the active element in the most common type of LDR.

COILS:

Simply a number of 'turns' of a conductor (usually insulated copper wire) to increase a property called 'inductance' (see technical terms).
The coil's characteristics will change depending on how many turns it has, how close together they are, their

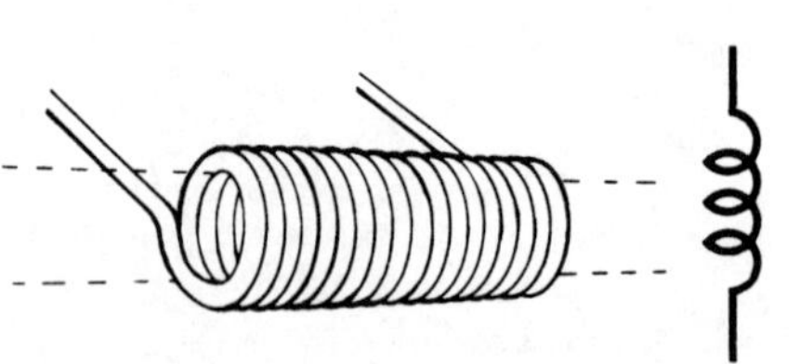

diameter and what they are wound on. For example, a number of turns on a wooden dowel will not have nearly as much inductance as the same number of turns wound around a ferrite rod of similar diameter.

FERRITE ROD AERIAL:
Small transistor radio type, two coils.
(Dick Smith Cat L-0520)

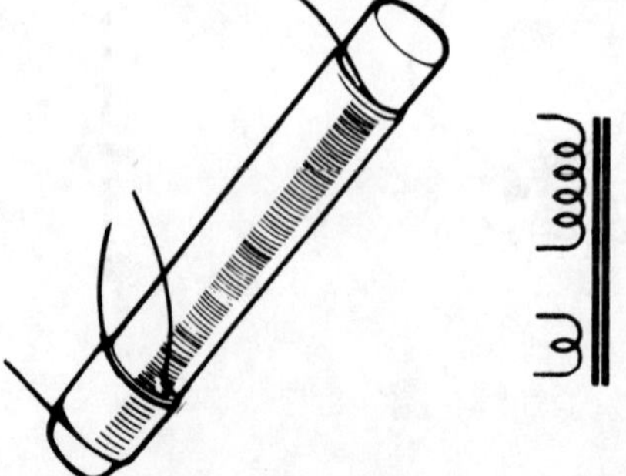

The ferrite rod aerial concentrates radio waves, in some cases eliminating the need for a separate aerial.
Normally with two coils, typically they are coloured red and pink. However, this need not always be the case. The wire in these is very fine: care must be taken to avoid breaking it.

TRANSFORMERS:

The transformer transfers AC signals from one section of a circuit to another. At the same time it can also isolate DC (Direct Current) from the rest of the circuit.

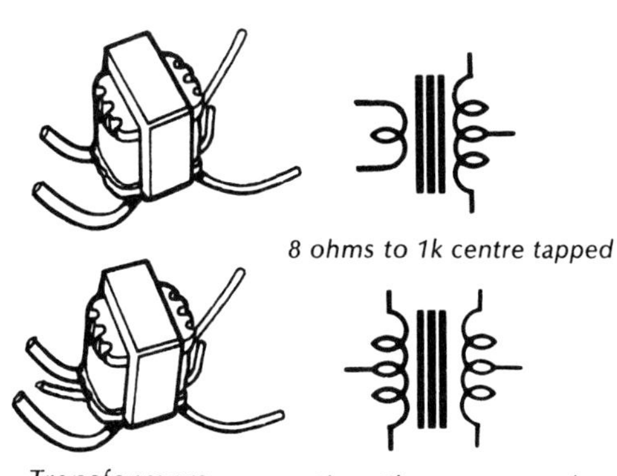

Audio Matching Types,
8 ohms to 1k centre tapped
(Dick Smith Cat M-0216)
3k to 3k centre tapped
(Dick Smith Cat M-0222)

Some transformers have a 'centre tap', which may or may not be used.

ELECTROMECHANICAL DEVICES

A component that changes an electrical state with a change in physical state can be grouped under this heading. Switches in one form or another are about the simplest example of this action. In this case, a lever or toggle can change the state of an electrical contact system.

The reverse of this action can also be applied, that is, where the electrical state produces a mechanical action. A solenoid or relay would be an example. Sound producing components, like speakers would also fit into this category.

RELAYS:

(Dick Smith Cat S-7120)

These are simply switches that are operated by an electromagnet instead of mechanical action from a hand or lever movement. As with other switches, the contact arrangement can be from a simple single contact to a multipole changeover system. The coil of the electromagnet can be powered by varying sources. The operating characteristics of this coil are usually refered to in volts, resistance and current required to 'pull in' the contact system.

The relay has two useful characteristics. Firstly, the coil requires low power with respect to the amount of power that the contacts are capable of switching. Secondly, the coil and the contacts are electrically isolated from each other (within limits). This means that two or more circuits of different potentials can be controlled by the same circuit action.

Relay

Single Pole Single Throw

Or

Push Button

SWITCHES

Any device (generally mechanical) capable of interrupting the flow of current through a circuit can be considered a switch. They vary greatly in shape and form depending on the end use requirements. The contact change action can be controlled by a lever, knob, toggle, slider or similar actuator. The number of contacts changed by this mechanical action may be anything from a single make/break or changeover to a multipole, multiposition design. The number of separate contact systems are referred to as the number of poles, the number of changes within a single system are referred to as the positions. For example, a switch haveing two contact sets with two different contact possibilities would be referred to as a double pole, double throw, abbreviated as DPDT.

Double Pole Double Throw

Double Pole Double Throw

Switches

MORSE KEY:

Economy Model
(Dick Smith Cat D-7105)

The contacts of this switch are generally simple in form being either a single pole, single throw (SPST) or a single pole double throw (SPDT). The specialty of this design is in the mechanics of the lever or key arrangement. It is structured in such a way that the movements of the hand require a minimum of effort and follow the natural rhythm of the wrist action. As the name suggests, it is used to 'key' or send Morse signals by switching on and off in the characteristic dot, dash pattern.

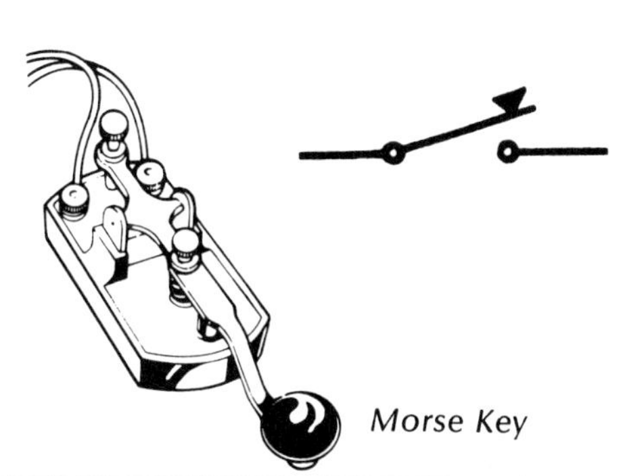

Morse Key

LOUDSPEAKERS

This component changes currents in the 'voice coil' to sound from a diaphragm, the 'cone'. The voice coil is simply an electromagnetic system within the confines of a strong magnetic field. The voice coil connected to the cone is allowed to move freely within the fields of magnetic force. As the magnet is fixed, any change in the current through the coil will cause the cone to move in one direction or the other depending on the polarity of the current flow.

A varying audio signal source, from a radio for example, will cause the cone to create pressure waves in the front and rear of it in sympathy with this signal.

A loudspeaker's voice coil is referred to as having an impedance and is stated in ohms. The amount of power the speaker is capable of handling is stated in Watts.

Speakers

MAGNETIC EARPHONE OR HEADPHONES

The magnetic earphone is similar in electrical action to the loudspeaker. Dynamic headphones manufactured today in most cases have miniature loudspeakers as the 'transducers'. These devices, like loudspeakers, have an impedance stated in ohms.

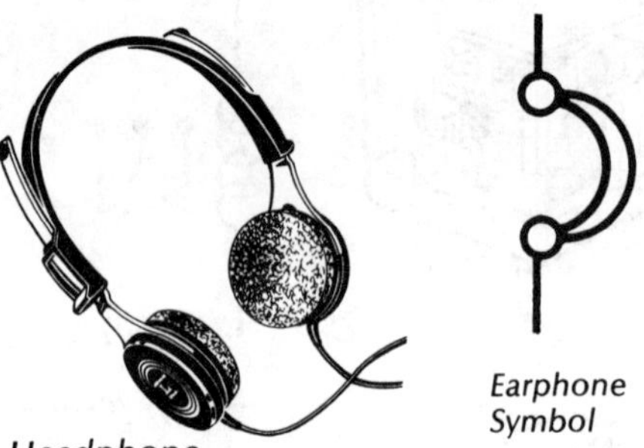

Earphone

CRYSTAL OR PIEZO EARPHONES AND HEADPHONES

With these components, the electrical energy is converted to mechanical energy as sound by the action of a crystal or piezo element on a diaphragm.

In this way, any change in electrical state across the element will result as sound.

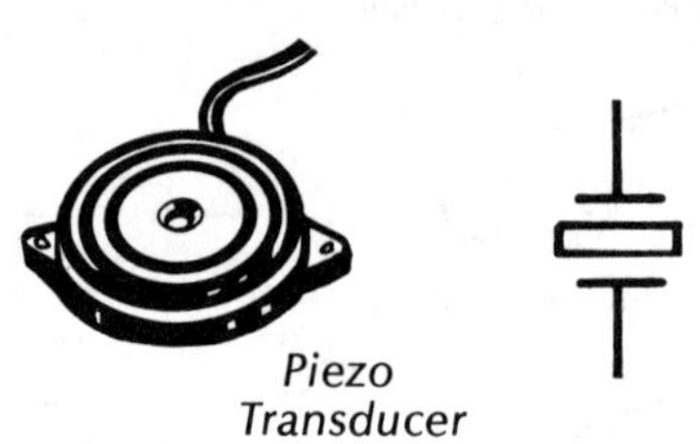

Headphone

Earphone Symbol

PIEZO TRANSDUCERS

These are sound transducers driven by an alternating voltage or pulse source. The element of the device is a flat diaphram with the piezo disc bonded to the surface. The whole assembly is fitted into a plastic case that is hollow inside with a single hole to the outside surface. This chamber is roughly tuned to the natural resonant frequency of the diaphragm. This frequency can be somewhere between 1kHz and 5kHz. The maximum output from these transducers can be expected at this fundamental frequency. The impedance of these devices may be only a few hundred ohms at resonance but increases to a very high value off frequency.

Piezo Transducer

MICROPHONES

There are basically four types of microphones – carbon, crystal, dynamic and electret.

Probably the oldest type of all, the carbon microphone has been used extensively in telephone systems. The insert is made up of carbon granules between two contacts. One of these contacts is a diaphragm which can be moved by sound waves. When it so moved, the granules are squeezed together, acting as a variable resistor that changes in resistance in sympathy with the sound.

The crystal type uses a piezo element similar to that of the earphone except in this case it is used in reverse. That is, the diaphragm and the element connected to it are stressed as sound pressure waves bombard the surface.
The impedance of this microphone is very high.

The dynamic type uses the speaker principle in reverse. When sound pressure waves move the diaphragm, the voice coil cuts the magnetic lines of force and therefore generates a small voltage. The impedance and output of these units is low but they often have an inbuilt transformer to increase the value to a generally more usable level.

Electret microphones are in effect a small capacitor where one of the plates is connected to a diaphragm, and a small electret charge is built permanently into the plastic dielectric material between the plates. When sound pressure waves move the diaphragm, the voltage across the capacitor varies to produce an output. A small amplifier is often built into this type of microphone, powered from a single 1.5V cell. In a case where only the insert is used, a third wire may be involved. This wire is a power supply connection to the internal amplifier.

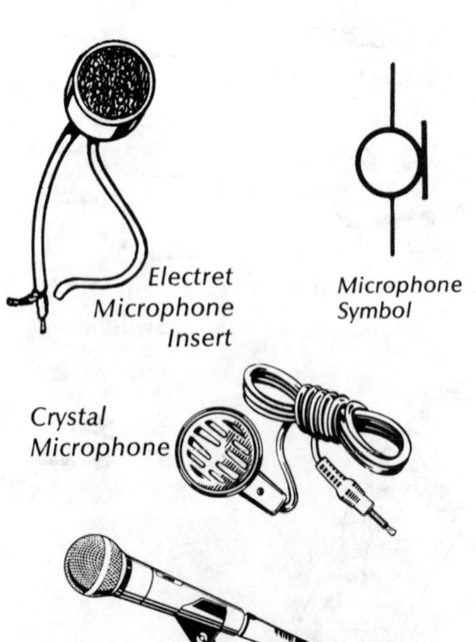

Electret Microphone Insert

Microphone Symbol

Crystal Microphone

Dynamic Microphone

TRANSISTORS:

The word 'Transistor' is actually another acronym – it stands for **Transfer-Resistor.** The original (and still most common) transistor type is the **Bipolar** transistor (see technical terms). They come in two versions, PNP and NPN.

There are generally three leads on a transistor and each one must be connected in the circuit correctly in order for the transistor to operate. One lead is called the **base**, sometimes abbreviated in a schematic circuit diagram as 'B'; the next lead is called the **collector** abbreviated 'C' and the other lead is called the **emitter** abbreviated 'E'.

Transistors are used as either switches or amplifiers in this book.

NPN TRANSISTOR:

BC548/DS548 or 2N2222
(Dick Smith Cat Z-1308)

An NPN has a positive voltage on its collector and a negative voltage on its emitter. When a positive voltage (with respect to the emitter) is applied to the base the transistor begins to conduct by allowing current to flow through the collector/emitter circuit. The relatively small current flowing through the base circuit causes a much greater current to flow through the emitter/collector circuit. This phenomenon is called current gain. It is a measure of the transistor's ability to **amplify.**

Note that the arrow in the diagram for the emitter points **out**.

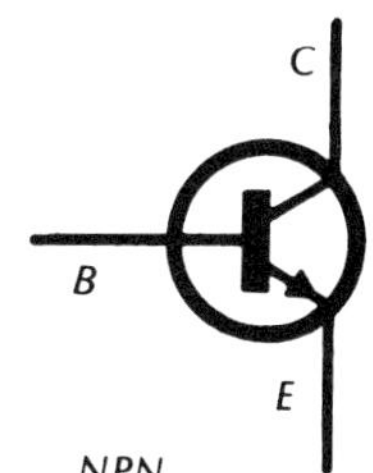

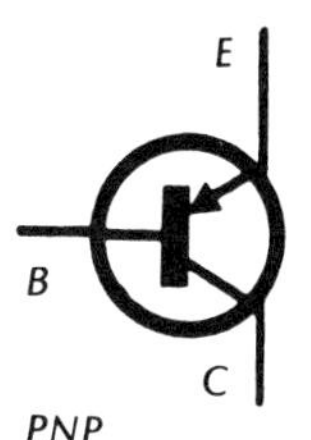

PNP TRANSISTOR:

BC558/DS558 or 2N3906
(Dick Smith Cat Z-1348)

A PNP transistor does **exactly the same thing** as an NPN except that it has a negative voltage on its collector and a positive voltage on its emitter. When a negative voltage (with respect to the emitter) is applied to the base, current will flow through the collector/emitter circuit. This time the current will flow in the opposite direction of course.

You can instantly recognise which type of transistor you are dealing with in a circuit by the fact that the arrow points outwards on an NPN and inwards on a PNP. The transistor may be drawn upside down, back to front, or any way in a circuit – but the arrows will still point out for an NPN, in for a PNP.

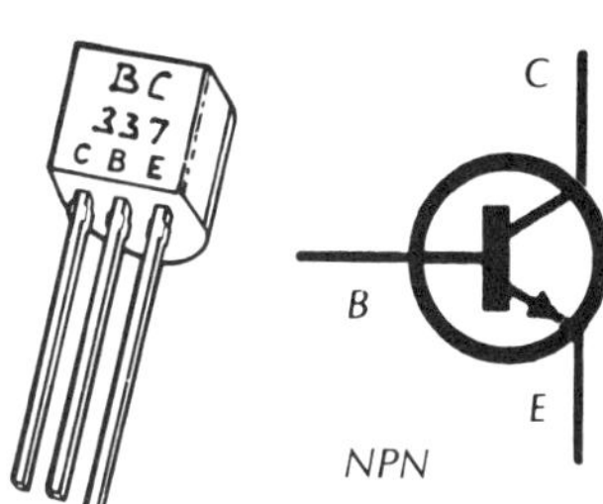

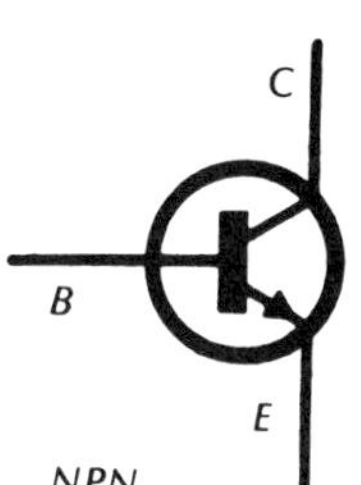

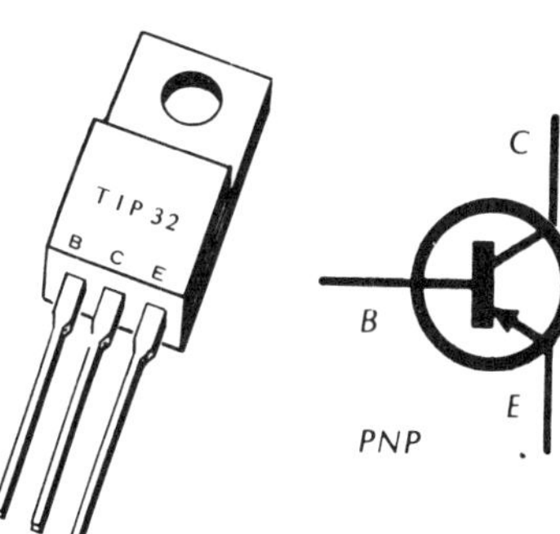

DIODES:

There are a number of different types of diodes, all with their various differences but with one basic feature in common: they let current flow in one direction only. Diodes are therefore, polarised and must be placed correctly in circuits or they will not work. If wired in the wrong way around they can be damaged, as well as causing damage to other parts of the circuit. The 'one way' feature of the diode is used in various ways and you will be introduced to several of these in this book.

POWER DIODES:

A typical diode with a band at one end to indicate the cathode (K).

Diodes are frequently used as 'rectifiers' to change alternating current to pulsating direct current, or with smoothing to direct current. Generally diodes for this use are known as POWER DIODES, such as type IN4002 or similar (Dick Smith Cat Z-3202). As seen in FUN WAY INTO ELECTRONICS VOL 1, these may be used simply as a reverse polarity protection device to protect the circuit in case the battery is inadvertently connected the wrong way round.

SIGNAL DIODES:

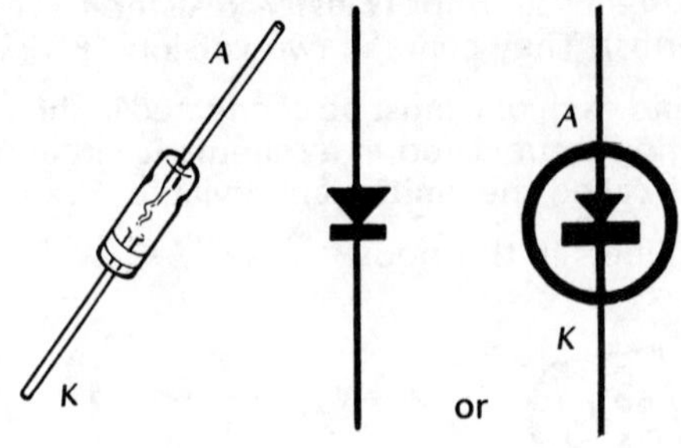

Diodes are also used in the detection circuits of some radio receivers, in which case they need to have a very low voltage drop. These diodes are called 'SIGNAL DIODES' (type OA91 or similar, Dick Smith Cat Z-4030).

ZENER DIODES:

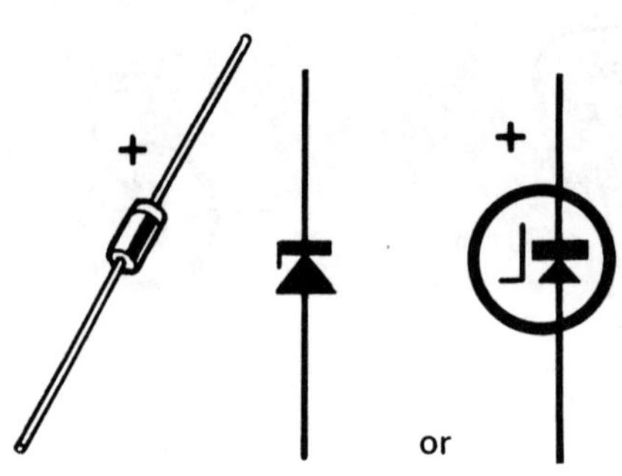

ZENER DIODES (Cat Z-3413 or similar depending on value). As with a normal diode, the zener diode conducts in one direction and blocks current in the other direction. But the zener diode blocks current only up to a certain voltage, when the reverse resistance drops to a low value and the diode conducts, in the normally reverse direction. When this occurs, the voltage drop of the diode remains almost constant, over a wide range of currents, so the zener diode can be used to 'clamp' the maximum voltage that can occur in a circuit. The voltage of a zener diode may be preselected, and zener diodes are sold according to voltage. So you can buy a 6.2 volt 'zener', or a 12V zener, and so on.

LIGHT EMITTING DIODES:
Red
(Dick Smith Cat Z-4010)

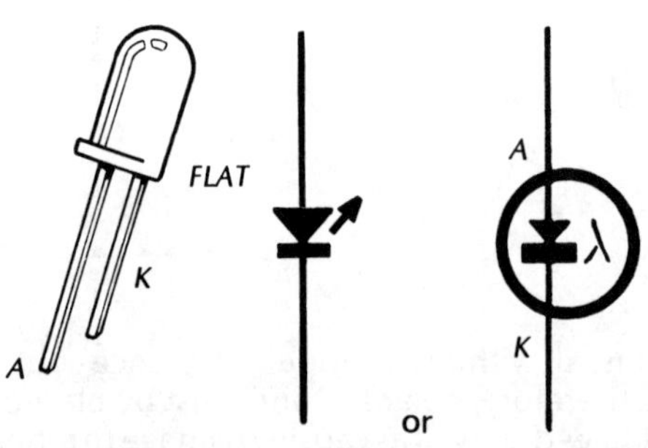

Called LEDs for short, these do get upset with reverse polarity. When correctly connected, they glow brightly. Their polarity is shown in two ways: they normally have a longer lead for the anode, and the cathode is often marked by a slight flattening on the body of the LED adjacent to it. They are also available in other colours and sizes.

LED DISPLAY:
EG. LT 303/313
(Dick Smith Cat Z-4103)
Common Cathode
LT302/312
(Dick Smith Cat Z-4102)
Common Anode

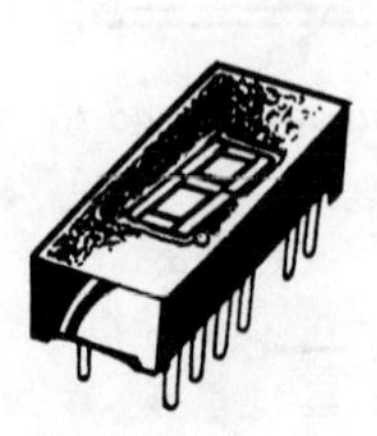

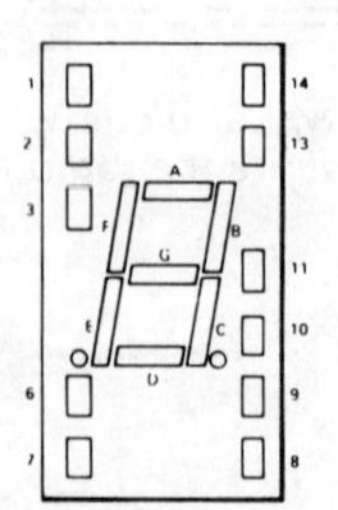

Basically a number of LEDs arranged in a particular way to create a special effect. For example project 9 uses a display which has 7 separate LEDs arranged in a figure-8 pattern. When power is applied to the display the numbers 0-9 can be obtained. This enables electrical signals to be directly converted to visual information. LEDs are used as displays for other purposes as well e.g. arrow signs, panel meters, etc.

INTEGRATED CIRCUITS

As the name implies, a circuit that is **integrated** onto (and into) a tiny chip of almost pure silicon. The level of complexity of 'Integrated Circuits' of IC's as they are commonly known, varies enormously.

A simple 'IC' can consist of just two transistors on the same chip of silicon. A more complex one can have tens of thousands of transistors on the one chip. Technology in this area is improving all the time and as you read this line new and more complex IC's are being produced. For the purposes of construction of this book, however, we only have to consider their function, i.e. what the whole IC **does** for us. Some IC's are complete amplifiers on a 'chip' whilst others can be computers on a 'chip'.

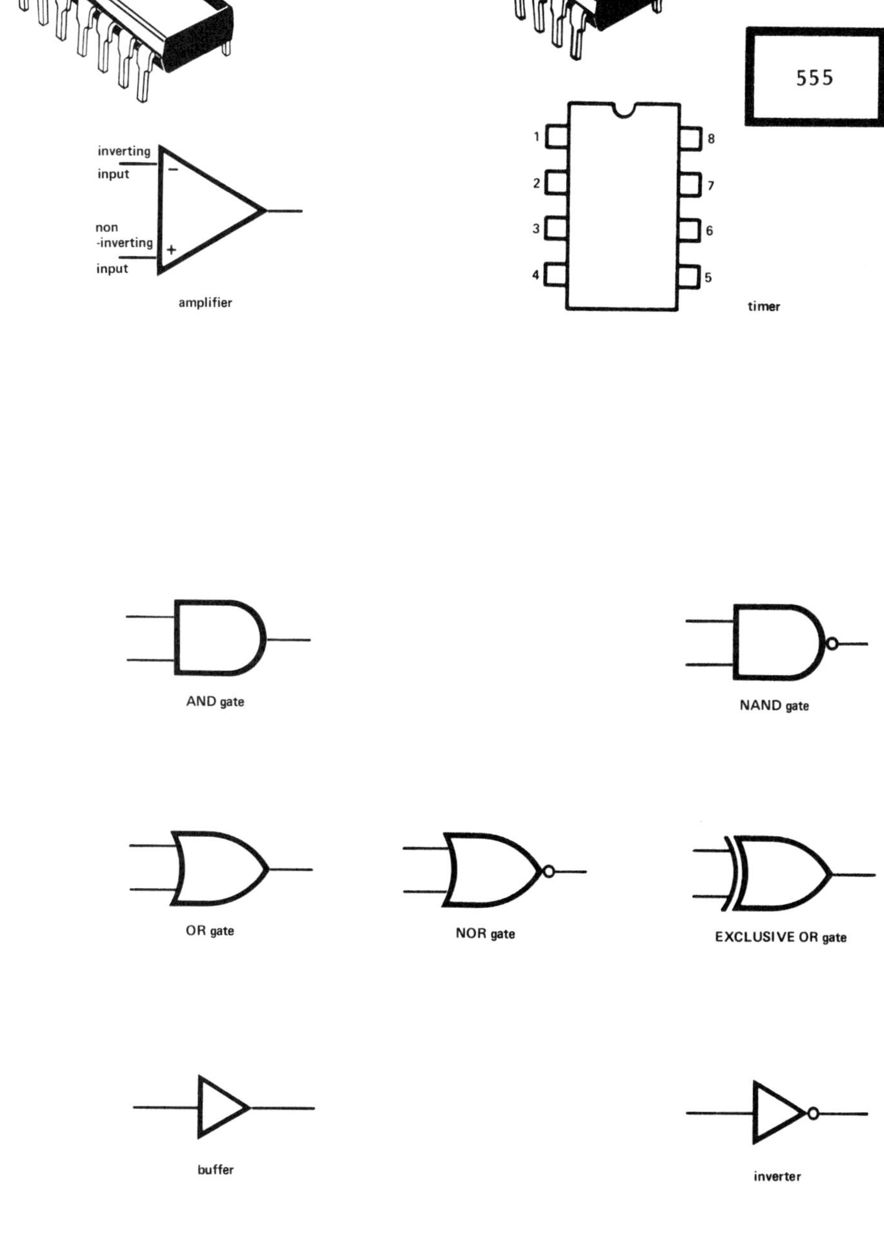

LINEAR INTEGRATED CIRCUITS

This is the general name given to IC's that perform functions of an analogue nature. The circuit behavior follows what could be loosely termed a 'linear' response (with exceptions) to signals. In a linear circuit, a signal at the input to the circuit appears at the output without modification except (normally) amplification.

Linear IC's can be made in various ways using different fabrication techniques but the basic functions remain the same. The internal structure of individual circuits may look complex to perform a specific job but the end result is the same. The diagrams of the linear devices used in the projects in this book are shown with the relative pinouts on pages 18 and 19.

DIGITAL INTEGRATED CIRCUITS

Unlike linear devices, this IC group depends more on a fixed value state rather than a linear change. The state may be either high or low, represented as '1' and '0' respectively. The transition period between these two states is usually very fast and not relative to the whole cycle of events (with exceptions) as with an analogue circuit.

Common to this group is the circuitry of logic systems where these two states represents binary numbers (or digits), the basis of all digital computers.

Some of the digital circuits used in the projects in this book are based on what is commonly refered to as CMOS (Complementary Metal Oxide Semiconductor). This impressive sounding name simply refers to the manufacturing technique and the transistor types used in this IC. These devices operate over a wide voltage range but are generally not as rugged as their earlier TTL (Transistor Transistor Logic) counterparts. The diagrams on pages 18 and 19 show the pinout and basic circuit functions of the devices used in this project series.

HOOK-UP WIRE:

 As the name applies, hook-up wire is the general term given to wire that is used to connect various parts of a circuit. The copper tracks on the printed circuit boards do much of the 'hook-up' work these days. This is not always practical, however as some components (speakers for example) often can't be mounted on the PCB. In such a case you would use two pieces of hook-up wire to connect between the speaker and the PCB.

Hook-up wire is generally made up of many strands of fine wire (to aid flexibility), covered with insulating plastic. The thickness of hook-up wire depends on how much current it is expected to carry.

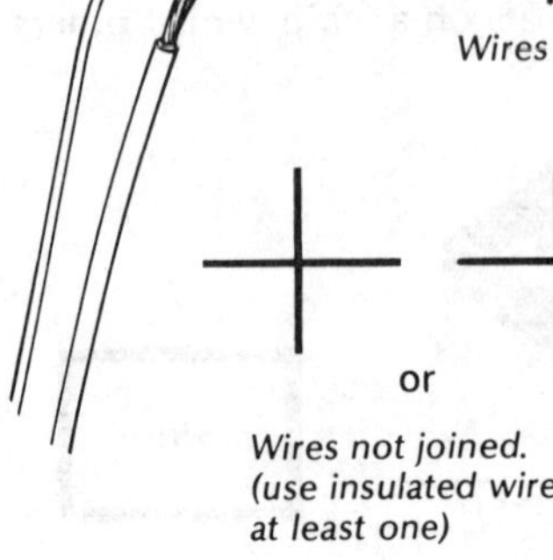

PRINTED CIRCUIT BOARDS:

 A flat piece of material around 1½ mm thick, generally made of phenolic or fibreglass material. On one side (but sometimes on both) is bonded a very thin sheet of copper. Parts of the copper are etched away to produce 'tracks', which form the conductors between components, which are soldered to them.

Called 'printed' circuit boards (or PCB's) because techniques similar to printing are used in their manufacture.

SPAGHETTI INSULATION:

Called 'spaghetti' because it is hollow like spaghetti, this is an insulator which may be used to slip over hook-wire to give more rigidity when used as 'probes'. Spaghetti insulation can also be used over tinned copper wire to insulate it if plastic covered wire is not available.

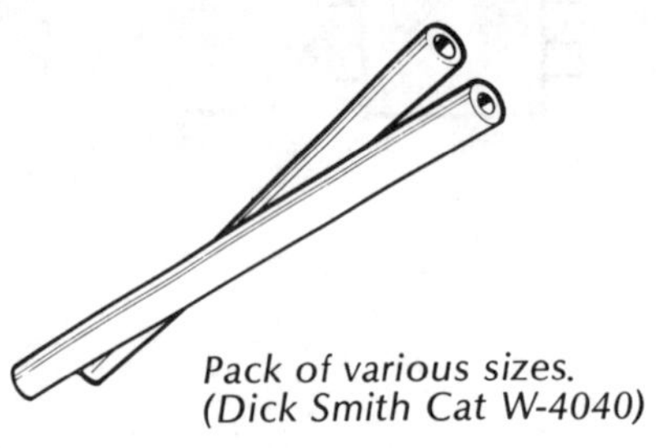

Pack of various sizes.
(Dick Smith Cat W-4040)

SOLDER:

 Broad name used to describe a material used for electrically bonding metals together. Usually made from an alloy of lead and tin, sometimes with traces of other metals. While common solder is 50% lead and 50% tin, solder used in most electronic work is generally 60% tin and 40% lead (or '60/40' solder). Electronic solder usually has a fine thread of 'flux' running down the centre of the solder wire. This saves a separate fluxing operation and makes a neater, more reliable joint. When constructing the projects in this book, you should use fine solder (either 18 or 22 gauge) for best results.

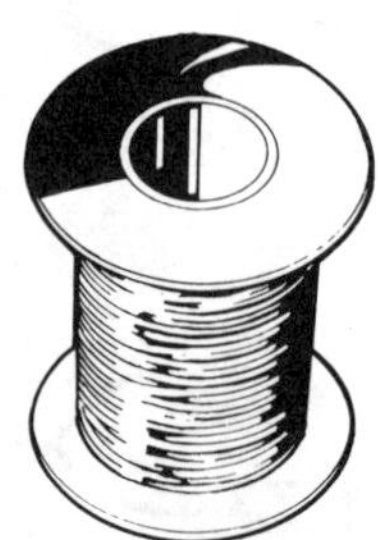

'PLUG-PACK' POWER SUPPLY: Dick Smith Cat M-9526

Strictly not a single component but a 'power supply'. They resemble a 3 pin mains plug except that they have a square body. Inside the body is a small transformer and other components which enable you to power your projects from the 240V mains, safely and cheaply.

You can regard a 'plug pack' as a power supply component.

Cat M-9526
Plug Pack

Cat M-9530
Power Supply

CMOS HANDLING PRECAUTIONS

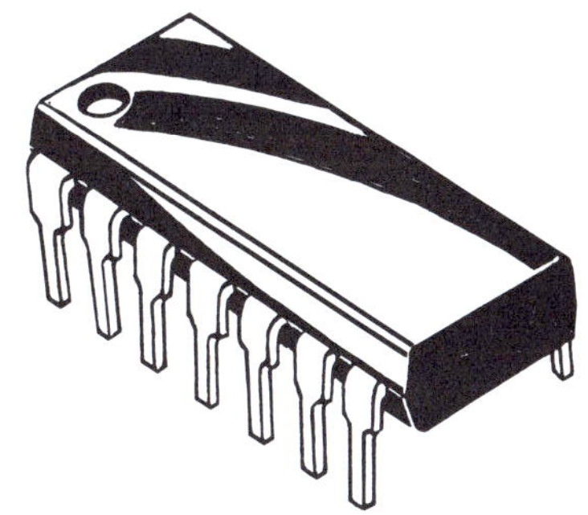

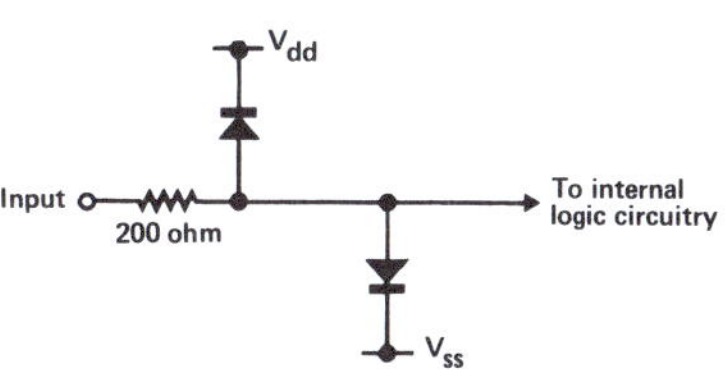

Typical input protection network for 4000 series CMOS devices.

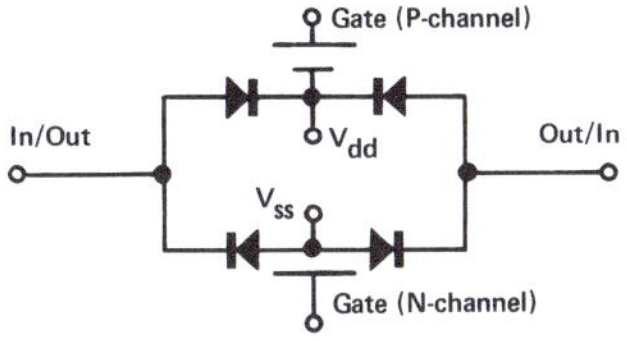

Transmission gate with intrinsic diodes to protect against static discharge.

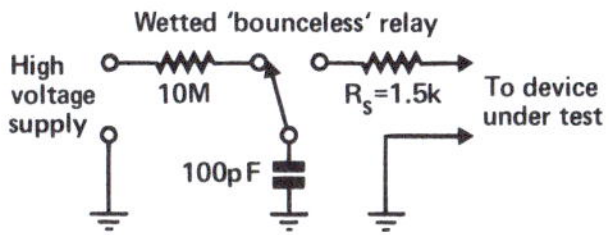

Equivalent RC network to simulate human body static charge, (Lenzlinger 2).

1. T.S Speakman, "A Model for the Failure of Bipolar Silicon Integrated Circuits Subjected to ESD", 12th Annual Proc. of Reliability Physics, 1974.

2. M. Lenzlinger, "Gate Protection of MIS Devices", IEEE Transac. on Electron Devices, ED-18, No. 4 April 1971.

Because of their high input impedance characteristics, CMOS devices (Complementary Metal Oxide Semiconductor) are subject to damage from high voltages applied to the inputs. These voltages could be from other circuitry sources, power supply fluctuations or from external static discharge.

All CMOS manufacturers incorporate some form of input stage protection but this may not always be sufficient in cases of severe electrical transient voltages from static discharge, abnormal operating or handling states.

Normal Operation

For normal operation, it is recommended that all inputs and outputs (unless specified) are kept within the constraints of the supply rail voltages Vdd or Vcc (positive rail), and Vss (negative rail). Before voltage can be applied to any input, power must be applied to the CMOS device.

No device should be removed or inserted from circuit or a socket with the power applied. Similarly, low impedance voltage sources such as pulse, waveform or function generator equipment should only be connected or removed when power is applied to the IC.

Power supply voltages should not exceed the normal ratings of CMOS devices; no greater than 18 volts between Vss and Vdd or Vcc (unless specified).

All unused inputs should be either tied to Vss or Vdd.

Condition	Most Common Reading (Volts)	Highest Reading (Volts)
Person walking across carpet.	12,000	39,000
Person walking across vinyl floor.	4,000	13,000
Person working at bench.	500	3,000
16-lead DIPs in plastic box	3,500	12,000
16-lead DIPs in plastic shipping tube.	500	3,000

Various Voltages Generated in 15% – 30% Relative Humidity (Speakman 1).

Handling Precautions

All devices not in circuit, (stored or transported) should be held in some form of antistatic material such as conductive foam, foil or the original manufacturer's aluminium or special antistatic plastic rails. They should not be stored unprotected in plastic bags or trays, styrofoam or similar static producing material.

High potential static voltages in the order of 4 to 15kV can be generated by a person walking on a waxed, carpeted or artificial floor. Woolen clothes brushing against a plastic or partial plastic chair can also produce high potentials. The human body may represent a storage capacitor of up to 300pF (with respect to ground) so this factor, along with the high voltage can deliver a high energy discharge into an object on touch. Most of us are aware of the phenomena and have been "shocked" at one stage or another with the characteristic "crack" that occurs at the point of discharge on touch.

Benches and Tools

Workbenches should be earthed if they have any metal parts. Tools, soldering irons and instruments should also be at ground potential.

In the case of amateurs, it is not always possible to use an earthed workbench but any wooden, felt, rubberized or similar surface is satisfactory. Avoid a plastic, Laminex or similar artificial surface. Plastic stools or chairs are not advisable.

Handling

Before you pick up a CMOS device, handle the circuit board or equipment on your workbench first. This in effect brings your body potential to the same as the component being handled. The CMOS device can now be picked up and inserted. This procedure is, of course, part of your normal movement so no problems should be experienced.

When a device is in circuit, the likelihood of damage from external static is very low provided that the inputs are terminated properly.

When a device is removed from circuit, if it is to be used again, it should be inserted into conductive foam or foil etc.

Integrated Circuit Pin Connections

The diagrams of integrated circuits shown on these pages are types used in the projects in this book. For full specifications refer to manufacturer's data.

CMOS – Digital

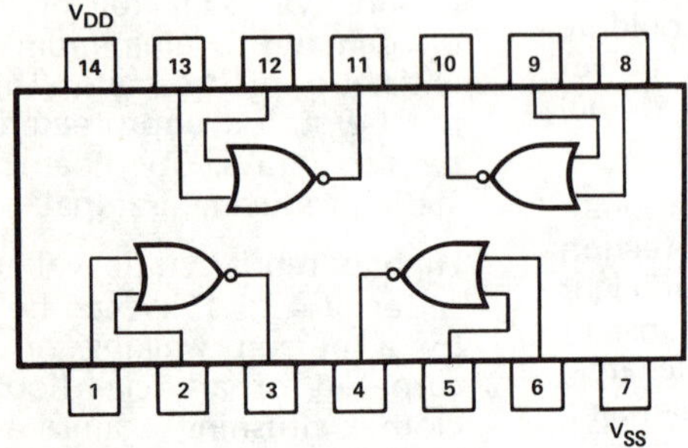

4001 Quad 2-Input NOR Gate

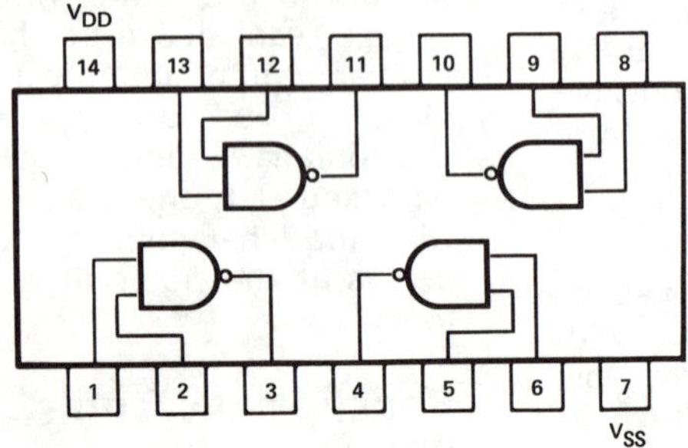

4011 Quad 2-Input NAND Gate

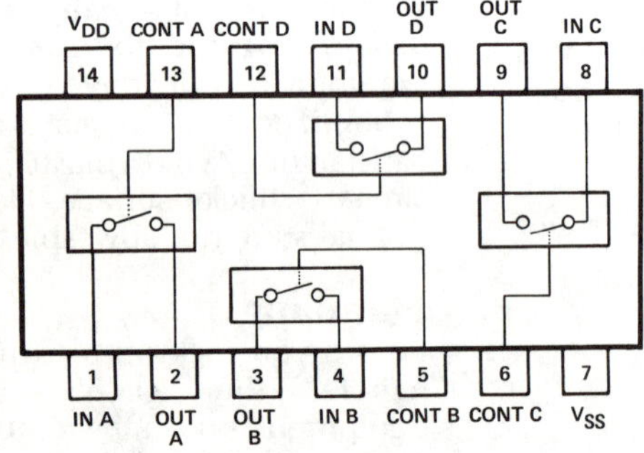

4016 Quad Bilateral Switch

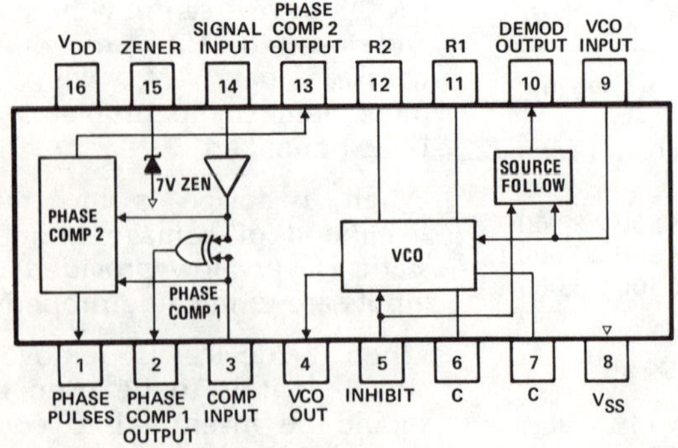

4046 Phase Lock Loop

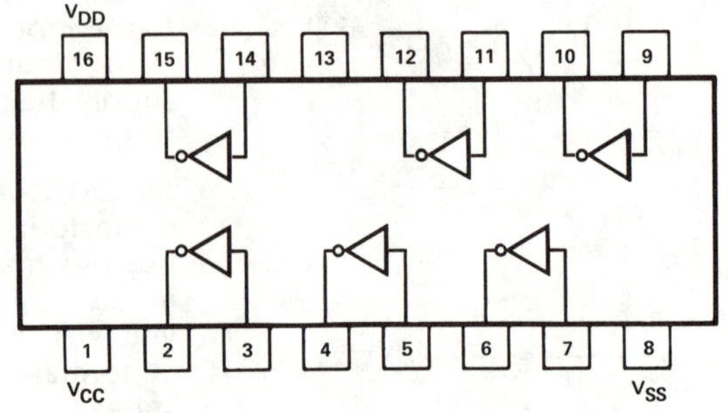

4009 Hex Inverting Buffer

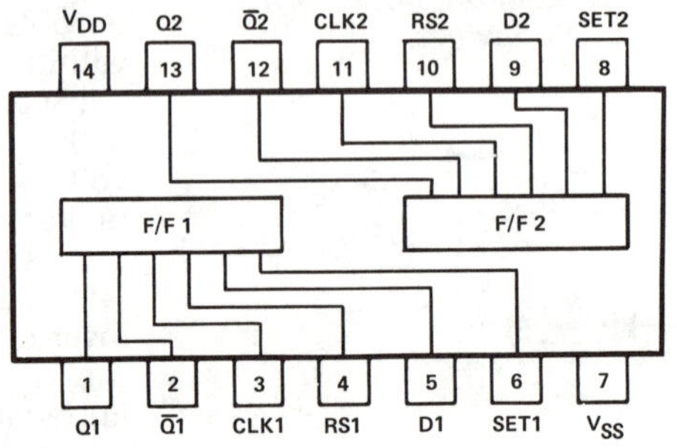

4013 Dual D-Type Flip Flop

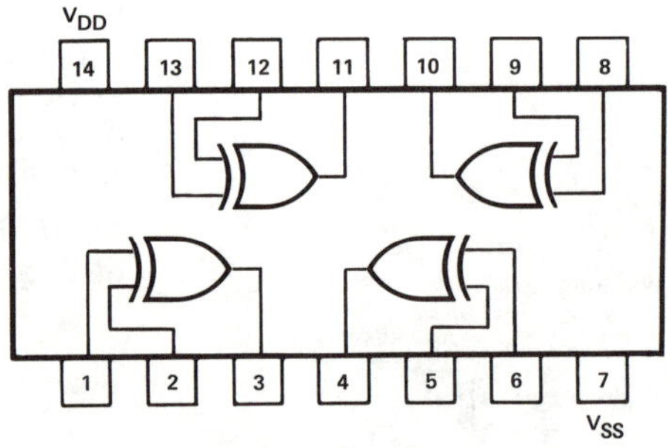

4030 Quad EXCLUSIVE OR Gate

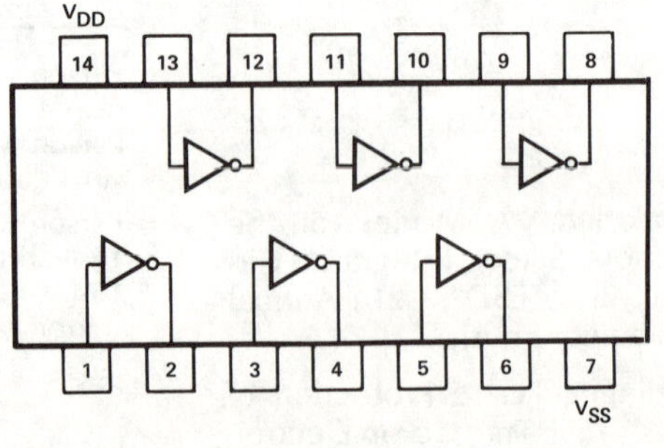

4069 Hex Inverter

CMOS (continued)

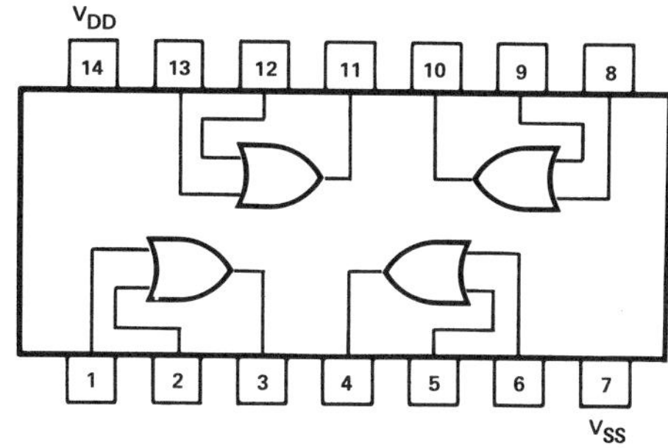

4071 Quad 2-Input OR Gate

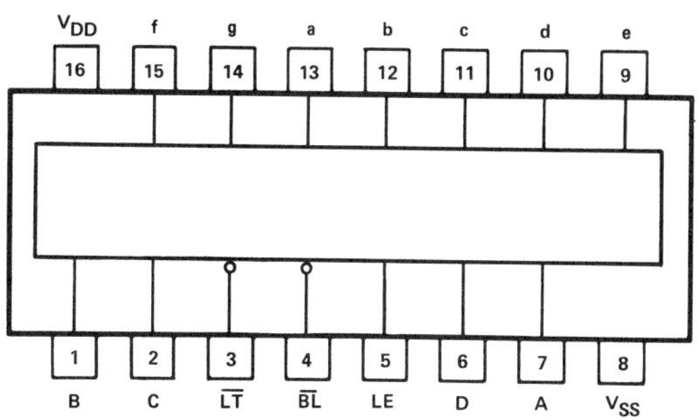

4511 BCD-to-7 Segment Latch Decoder/Driver

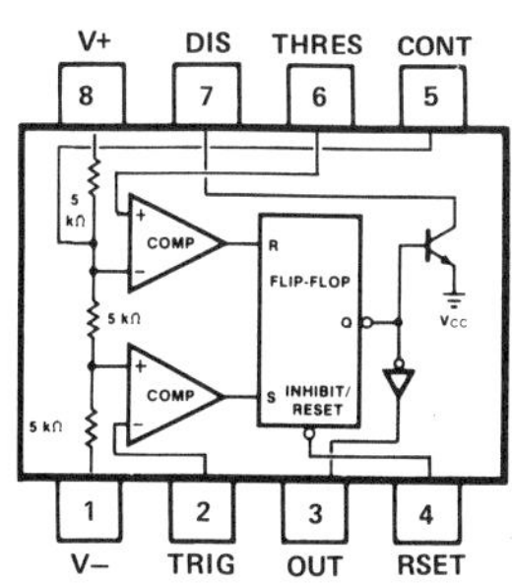

4518 Dual Synchronous BCD Counter

DISPLAY

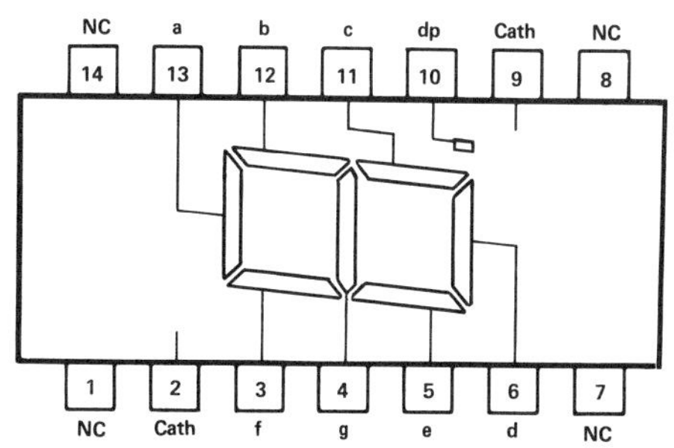

LT313 Common Cathode 7 Segment Display

LINEAR

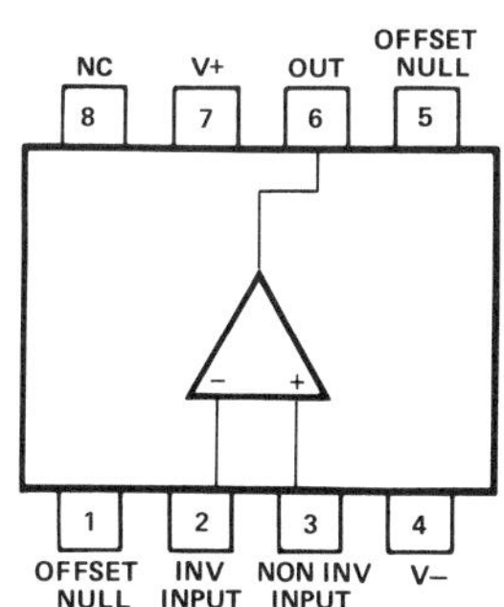

555 General Purpose Timer

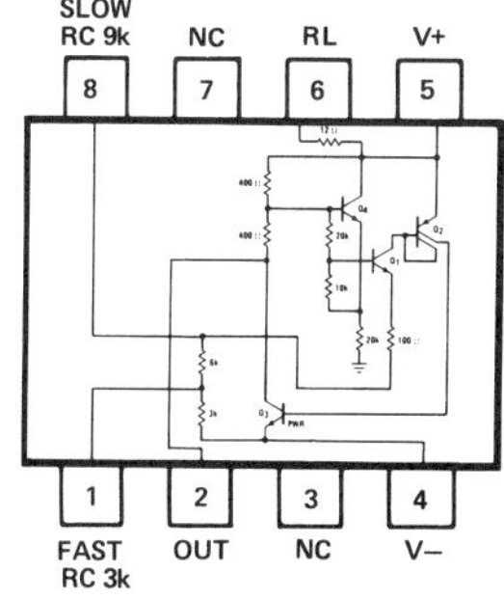

741 Operation Amplifier

LM3909 LED Flasher/Oscillator

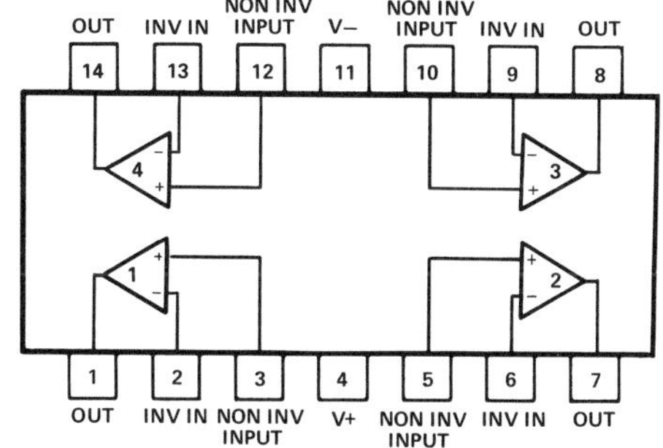

LM324 Quad Operational Amplifier

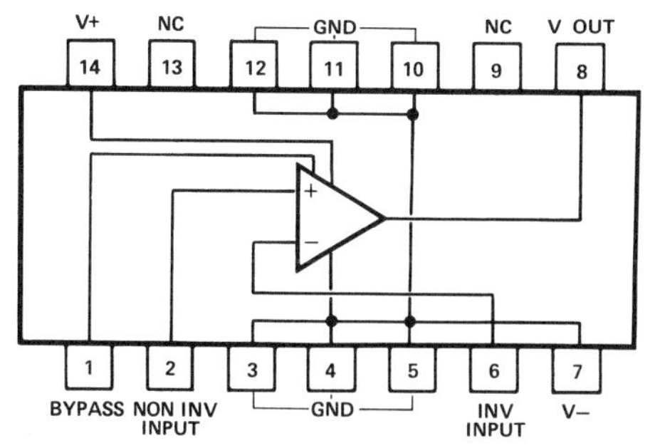

LM380 Audio Power Amplifier

Component marking codes

SEMICONDUCTORS:

Semiconductor devices (transistors, diodes, IC's, etc) come in an enormous variety of shapes and sizes. Despite this, there is a remarkable amount of standardization.

Although semiconductors may come from different manufacturers all over the world, provided that the device carries the same type number, the electrical specifications of that device should be the same. In some cases the Alpha prefix of a type number may be changed but the actual number is unchanged. This is common with transistors and the TO-92 case types shown below are typical. A DS549 is the same as BC549, a DS548 is the same as a BC548. In most cases, the pin connections are the same but if they are not physically marked on the body of the device or some doubt exists, refer to manufacturer or supplier data.

Transistors:

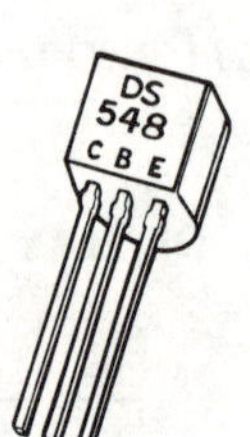

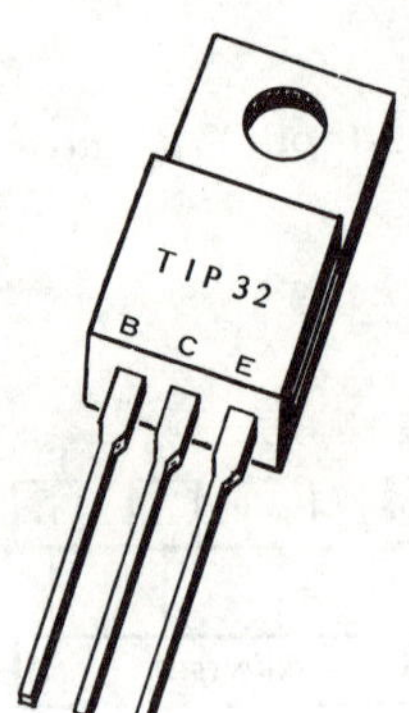
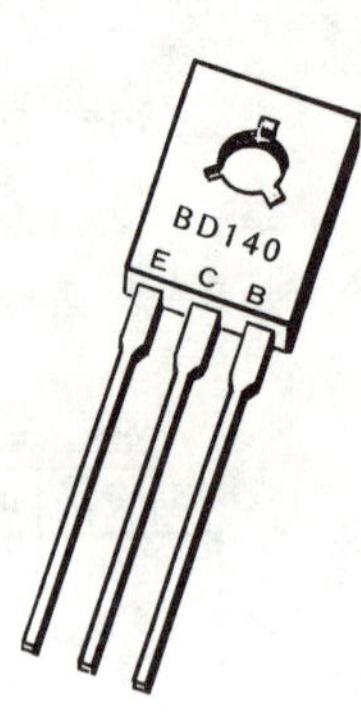

Diodes:

Usually have a band at one end indicating the cathode end.

A typical diode with a band at one end to indicate the cathode (K).

Circuit symbol for a diode.

IC's:

For the sake of this explanation we will only refer to the IC's used in this book.

These IC's are made in 'Dual-in-Line' packages or 'DIPs' in various lengths.

All DIP IC's in this book—indeed as far as we know most DIP IC's made, conform to the lead numbering system shown.

Viewed from the top (always) with the pins pointing away from you.

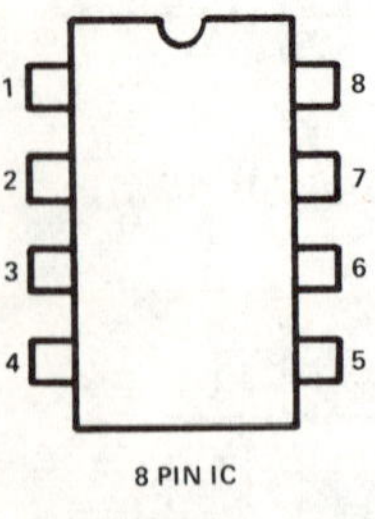

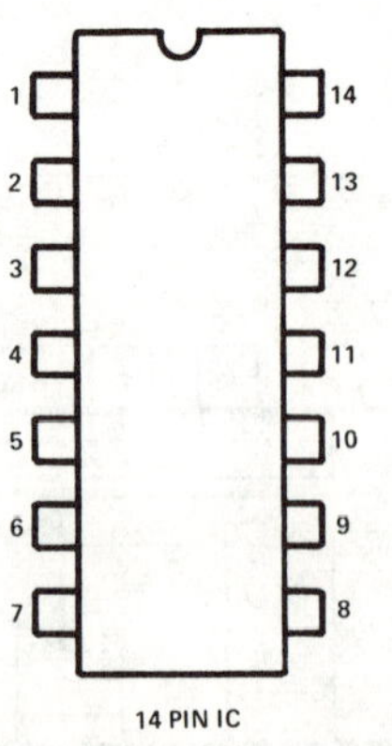

Index area. Within this approximate boundary you should find some sort of orientation mark. It is usually a notch in the plastic but it may be a spot of paint or a printed symbol etc.

Top left pin is always No.1 pin.
Pin 2 is directly below that and so forth.

Last pin.

Number sequence rotates in an ANTI-CLOCKWISE direction. Pin in bottom right hand corner is the next number after pin in bottom left hand corner.

General rule for IC's whatever number of pins.

REMEMBER: (i) These rules apply only if the IC is viewed from the top.

(ii) If in doubt, consult the manufacturer for information.

Resistors

Resistors are usually so small it is impractical to try and print each one with its value, so they are marked with a code printed on them in bands of different coloured paint. These bands give us the resistor's value.

The colours and their values:

BLACK	0	BLUE	6
BROWN	1	VIOLET	7
RED	2	GREY	8
ORANGE	3	WHITE	9
YELLOW	4	GOLD *	x0.1
GREEN	5	SILVER *	x0.01

** RARELY USED*

How to read them:

Start with the band closest to one end or, if it is difficult to work out, the band furthest away from the gold or silver band. Take the resistor shown to the right: let's say that the first band is orange—from the table this means three. If the second band is white, then its value is 9, from the table. And finally if the third band is brown we have a value of 1, so we add one zero to the first two figures: 3, 9 and 0—or 390 ohms.

The fourth, or tolerance band, is silver: 10%. Therefore, the resistor is 390 ohms, 10% tolerance (or somewhere between 390 - 10% and 390 + 10%, or 351 and 429 ohms). This is more than adequate for most circuit requirements. Resistors are available which are 'spot on' in value – however, these are not required in most circuits, and are often very, very expensive.

Close tolerance resistors are available that are marked with five colour bands. In this case, the first 3 bands are read off as digits, the fourth band is the multiplier and the fifth is the tolerance. Eg: Red, black, green, brown and brown would read 2050 ohms, 1%

Capacitors:

Most capacitors will have their value printed on them. However, there are a number of capacitor manufacturers who use the 'IEC' code. This code is a numerical code, but works in a similar way to the resistor colour code: two figures followed by a multiplier. There is often a single letter code showing tolerance.

The code is worked out in picofarads, so you may have to use the appropriate metric multiplier.

A capacitor may have a code '104K'. This decodes as follows – 0.1uF, 10%. The first two figures give us 10, the third figure gives us 0000, and the letter 10%. Therefore, the capacitor is 10 0000pF, 10%. We normally express this as 0.1uF – capacitors with a value below .001uF are expressed in picofarads.

There may be a further figure marked – this would be the voltage rating of the capacitor (the maximum voltage at which you can use that capacitor).

Metric units and conversions:

Throughout this book and indeed, most electronics publications, metric multipliers are used to simplify and shorten component values, etc.

For example, capacitance is measured in Farads—but the Farad is a huge unit of measurement, much too large to be of any use in expressing capacitor values. So we use a suitable metric sub-multiple to save a lot of figures. Instead of saying, for exmple, .000000000001 Farads we say 1 pico Farad (1 pF).

We have used only micro Farads (uF) and pico Farads (pF) in this book; however, some publications use the abbreviation nano Farad (nF) so you should get to know these eventually, too.

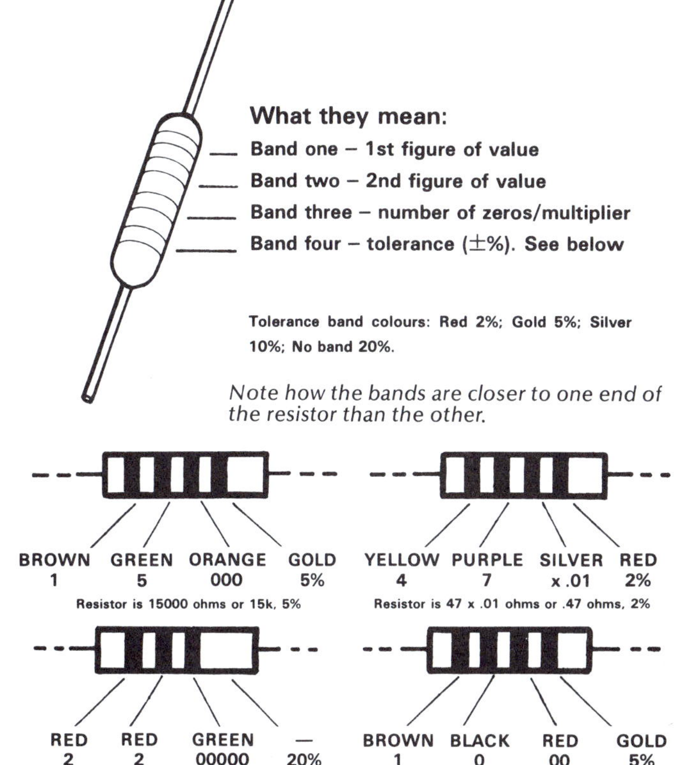

Note how the bands are closer to one end of the resistor than the other.

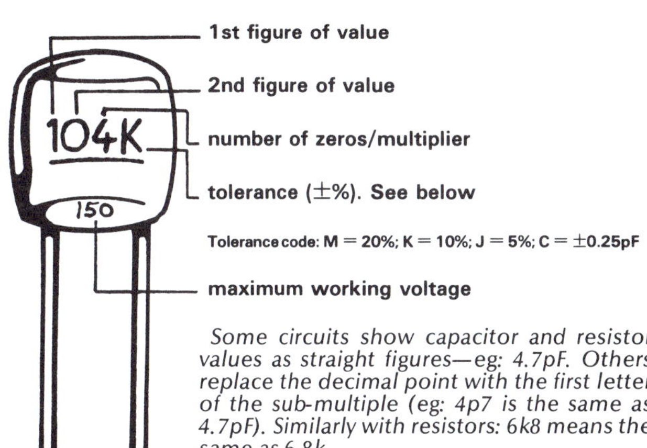

Some circuits show capacitor and resistor values as straight figures—eg: 4.7pF. Others replace the decimal point with the first letter of the sub-multiple (eg: 4p7 is the same as 4.7pF). Similarly with resistors: 6k8 means the same as 6.8k.

Abbreviation	Means	Multiply unit by	Or
p	pico	.000000000001	10^{-12}
n	nano	.000000001	10^{-9}
u	micro	.000001	10^{-6}
m	milli	.001	10^{-3}
—	UNIT	1	10^{0}
k	kilo	1 000	10^{3}
M	mega	1 000 000	10^{6}

1000 pico units = 1 nano unit	1000 nano units = 1 micro unit
1000 micro units = 1 milli unit	1000 milli units = 1 unit
1000 units = 1 kilo unit	1000 kilo units = 1 mega unit

Reading circuit diagrams

At first circuit diagrams will confuse you if you have not already gained some familiarity with them from FUN WAY INTO ELECTRONICS VOL.s 1 & 2.

It will help you to realise that the circuit symbols are simply the alphabet of another language. We all use several languages, in addition to the normally written and spoken English language. For example: road signs—used by motorists and pedestrians; the Greek alphabet—used in mathematical equations; company logos—designed to communicate a company's function, often without using words; the symbols in public buildings to designate various rooms. We learn to use these signs through familiarity gained by asking what they mean and by using them.

When you are learning the symbols used in electronic circuit diagrams, try and relate the symbol to the actual component it represents. For instance, a capacitor is really two conductive plates separated by a gap. A variable capacitor has an arrow on one plate, which is curved to show that it can be turned, or an arrow across both plates meaning the same thing.

The wriggly line of the resistor symbol immediately suggests a more difficult path for the flow of electrons, i.e., a more resistive path. Again an arrow indicates a variable resistor. Coils and transformers which in reality have coils of wire are shown as coils in the circuit diagram.

Diodes, which pass current in one direction and block it in the other, are represented by a symbol which shows an arrow in one direction and a line blocking flow in the other. Integrated circuits are shown either as a rectangular 'block', or much as they look with their polarity and pin numbers clearly indicated.

Using these examples as your guide, check through all of the components comparing the symbol with the diagram and with the actual component in each case. Soon you should be able to look at any circuit diagram and readily name each component.

Once you are able to recognise the symbols and understand what each component does, you are well on the way to understanding circuit diagrams. All that remains is to be able to read how the different components are wired together.

This becomes quite straightforward when you know that a small dot at the junction of two or more wires shows that they are joined, while a semicircle or hump indicates one wire passing over another without contacting it.

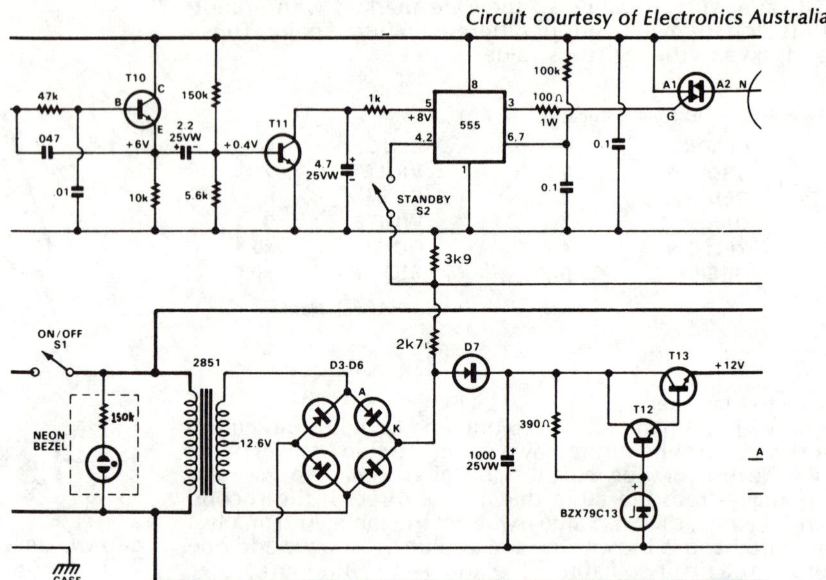

This circuit diagram is typical of those we use in this book. By now, you should be able to recognise many of the circuit elements: resistors, capacitors, diodes, transistors, etc. The strange symbol in the top right hand corner represents a 'triac'; a device used extensively in power control circuitry. The box-shaped symbol near it is for an integrated circuit; in this case, a 555 timer IC (we use this IC in some of our Fun Way projects).

As indicated in the section on components, some components are polarised—i.e., they have positive and negative symbols marked on them. These components must always be connected into the circuit with correct polarity, otherwise damage may occur to the component itself or to other parts of the circuit. If a polarised component is connected the wrong way around, generally the circuit will not work at all.

The symbols shown next to the component section are standard symbols used throughout the world (with minor variations) so even if a circuit appears in a foreign language handbook, you should still be able to interpret it.

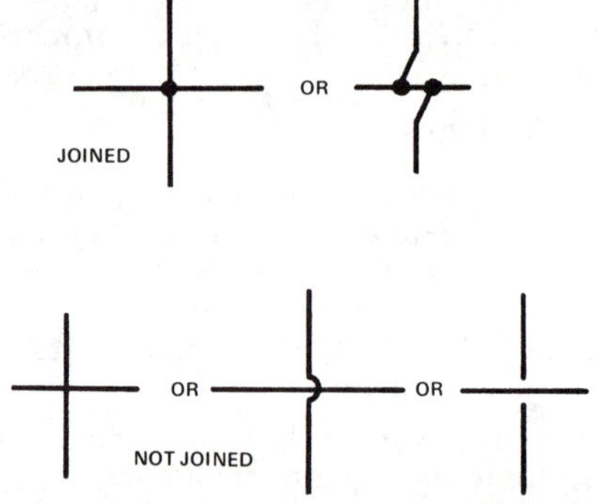

These two groups of drawings show how joins and cross-overs may be shown in circuit diagrams. In Fun Way 3, we use the first way shown on each line. However, other publications may use the alternatives.

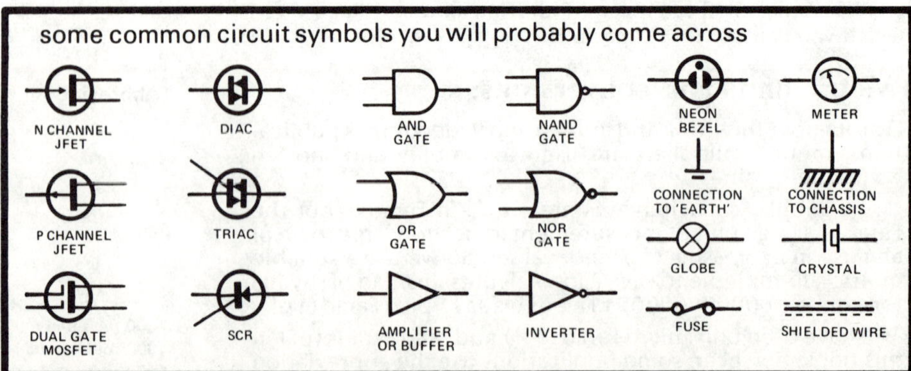

Circuit laws

It is not essential that you learn these laws to build the projects in this book, but they are very simple, and an understanding of them will increase your enjoyment of the subject. You may also use these laws to design your own circuits, find faults, etc, as your knowledge of electronics grows.

Ohms' Law

The relationship between voltage, current and resistance in a circuit is defined by Ohm's Law, which may be simply stated as: 'When a voltage is applied to a resistive circuit the current in Amperes will be proportional to the voltage and inversely porportional to the resistance in Ohms.'

This relationship is represented mathematically by the formula:

$$E = I \times R$$

where E is in volts, I is in amps and R is in ohms.

This can be turned around to look like:

$$I = \frac{E}{R} \qquad R = \frac{E}{I}$$

If you find these difficult to remember, you may find the diagram below helpful:

Simply cover the value you are looking for with your fingertip, and the formula you require remains exposed.

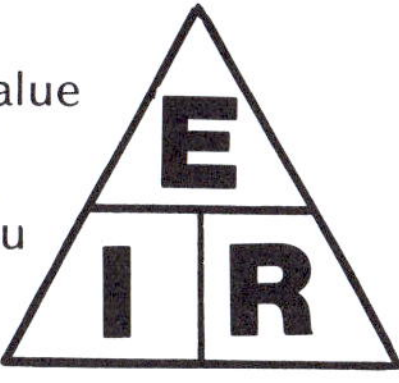

For example:
In a circuit consisting of a 6V battery and a 12 ohm resistor, we wish to find the current passing through the resistor.

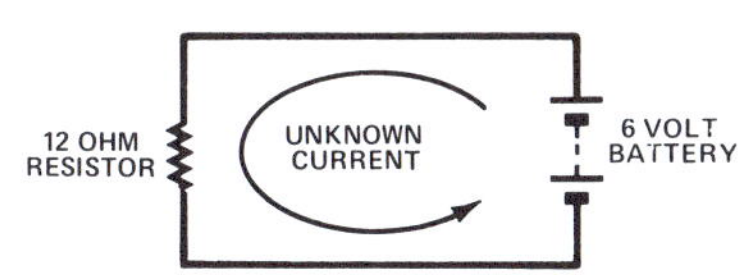

Because we want to know 'I', cover it with your fingertip. The formula to find I is then shown as E divided by R. Substituting values, we know E is 6 volts and R is 12 ohms; therefore the current is 6 divided by 12, or 0.5 Amps.

Suppose we knew the voltage and the current (we could measure these with a multimeter) but the resistor value has been rubbed off. Using the same triangle above, we cover R and find the formula R equals E divided by I. Therefore, R equals 6 divided by 0.5—or 12 ohms. It agrees with the above answer!

Now try one for yourself. We know I is 0.5A, and we know the resistor is 12 ohms. But we're not sure about the battery voltage. How do you work it out?

Power in a circuit

When a current passes through a component, energy is given off in the form of heat. Normally, we associate resistors with this action: that's part of their job. We often need to know how much power is being given off by a resistor—and we find this out by using a formula derived from Ohm's Law. This formula says that power dissipated is equal to the voltage across the component multiplied by the current flowing through it; or

$$W = E \times I$$

where W is in watts, E is in volts and I is in amps. This formula, too, can be turned around if required:

$$E = \frac{W}{I} \quad \text{or} \quad I = \frac{W}{E}$$

For example:
In the circuit above, E is 6 volts and I is 0.5 amps. Therefore the power dissipated would be 6 x 0.5 or 3 watts. We would have to use a resistor capable of dissipating at least 3 watts—we would probably use a 5W type.

Series and Parallel Circuits

In all circuits, combinations of components are used to achieve various effects. It is often essential to be able to work out the equivalent values of components connected together. To do this, one must be able to work out whether the components are connected in a 'series' or 'parallel'—or a combination fo both.

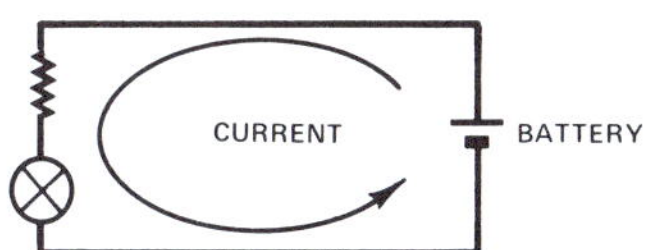

In a series circuit, current flowing from the battery must pass through all the components. Because of this, the current is the same through all components.

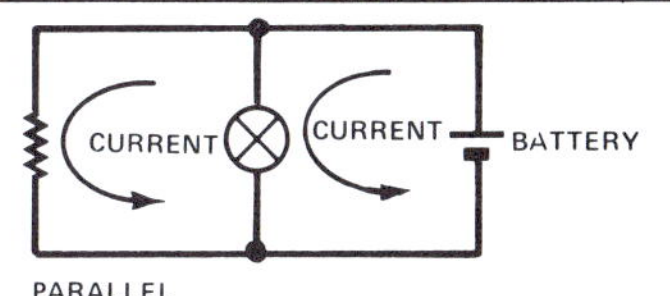

In a parallel circuit, the current can take a number of different paths — so currents are not identical through various 'legs' of the circuit.

Resistors in series

Resistors in a series circuit are simply added together to find the total resistance. In other words, a 10 ohm, 150 ohm and 1000 ohm resistor connected in series would be the equivalent of a single 1160 ohm resistor. The formula is:

$$R_T = R_1 + R_2 + R_3 + \ldots$$

Resistors in parallel

Resistors in a parallel circuit are a little more difficult. Here the reciprocals are added together to give the reciprocal of the total. The formula to use is:

$$\frac{1}{R_T} = \cfrac{1}{\frac{1}{R_1} + \frac{1}{R_2} + \frac{1}{R_3} + \ldots}$$

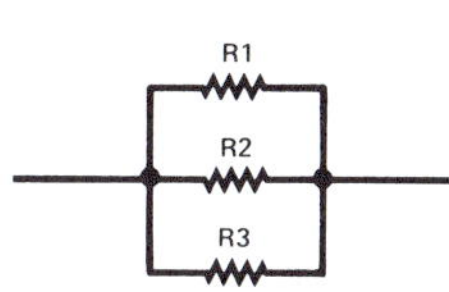

Capacitors in parallel

Capacitors behave exactly the opposite to resistors: when capacitors are in parallel, you add them:

$$C_T = C_1 + C_2 + C_3 + \ldots$$

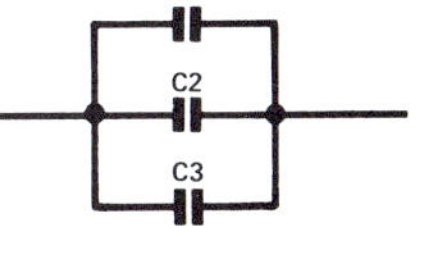

Capacitors in series

Capacitors in series, on the other hand, are similar to resistors in parallel—you add the reciprocals:

$$\frac{1}{C_T} = \cfrac{1}{\frac{1}{C_1} + \frac{1}{C_2} + \frac{1}{C_3} + \ldots}$$

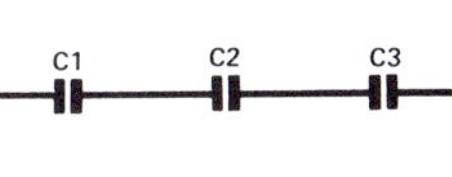

Two resistors in parallel or capacitors in series: A much simpler formula can be used if there are only two resistors in parallel or capacitors in series. It is: (For capacitors simply replace R with C).

$$R_T = \frac{R_1 \times R_2}{R_1 + R_2}$$

How to use your multimeter

So you've just arrived home from your local Dick Smith store with your brand new shiny multimeter. All you need now is to work out how to use it!

Unpacking it and setting it up:

When you buy a multimeter, it is seldom ready for use. Almost always, batteries must be fitted, and in some cases, there is a protective wire link across the meter to remove.

Read the instruction manual carefully, and carry out any steps described. A word of caution: most meters are supplied with batteries, but sometimes these are not the best. As a meter is left unused for a long period, it is a wise investment to buy some fresh, well-known-brand batteries.

Your multi-meter will have a large scale area, a knob surrounded by figures, a small knob marked 'Ohms adjustment' or similar, and at least two terminals. If there are more than two terminals, look for the ones marked '+' and '−', or 'V-A-OHMS' and 'common'. In many meters, these terminals will be coloured red and black respecitvely.

Plug your test leads into the terminals: red into the red or '+' terminal, black into the black or '−' terminal. Turn the knob to one of the 'ohms' ranges and short the ends of the test leads (or 'probes') together. The pointer should move at least part way up the dial, perhaps right to the end or past it.

If it does, your meter is ready for work. Read on!

What scales have you?

Most multimeters (even the real budget models) give you at least four different measurement options: DC voltage, DC current, AC voltage and Ohms. These are further divided into different ranges, to give more flexibility. The more expensive meters may also offer AC current measurement, and possibly other functions too.

Each of the ranges is represented by a scale on the meter face; though many of the scales may be shared. For example, you might read 0 − 2.5mA on the same scale as you read 0 − 250mA and 0 − 250 volts. These scales are linear; thus the divisions betwen markings remain constant. All you do is apply the right 'units' to the scales to work out the exact figure: these units are given to you alongside the pointer on the 'range' switch.

There might also be a strip of mirror on the scales. The purpose of this is to allow you to line up the pointer with its reflection, eliminating 'parallax' errors (errors caused by viewing the pointer from an angle).

There may also be other ranges marked on the scale, which are obtained by using different terminals to the normal ones. An example of these are on the meter at right: a 0.1V (50uA) DC terminal and a 1000V AC terminal, which are not affected by the 'range' switch. These special terminals are connected direct to the meter in some cases, and must be used with extreme care.

Measuring DC Current:

The important thing to remember when measuring current is that the meter must be placed **in series** with the circuit. This means that the circuit must be broken, with the meter leads connecting the two sections together.

Normally, a circuit is broken by unsoldering a wire or component; unfortunately, this is not always possible. For instance, if you want to know the current flowing through a certain PCB track, you must break the track (shudder!) to include the meter in series. This can usually be done by cutting the track at a narrow point with a very sharp blade. When the measurement is finished, solder can be flowed over the knife cut to restore connection. (You might even need to put in a small piece of hookup wire to assist the connection).

When measuring current, start with your meter on the highest range and work down. Stop when the pointer moves reasonably high up the scale.

It is also important that the probes are connected in the correct polarity. The red probe is connected to the more positive side of the circuit; the black probe is connected to the more negative side.

If you find the pointer swings the wrong way, reverse the probes.

Measuring AC Current:

Your meter may not have provision for AC current measurement, particularly if it is an economy type. If it does, measurement is basically the same as for DC: except there's no need to worry about polarity. Once again, start with a high range and work down to the correct one.

Measuring DC Voltages:

Measuring voltages is exactly the opposite to measuring current: the meter is placed **in parallel** to the circuit or component being measured. Once again, however, polarity has to be observed. Red (positive) to the positive side of the circuit, black (negative) to the negative.

And again, if you don't know the voltage, start on the highest range and work down.

Measuring AC Voltages:

It's just the same as measuring DC voltages: just follow the same steps and remember that the meter must be placed in parallel.

A word of explanation here about what the meter actually reads: As you know, an AC wave starts at zero, rises to a peak, falls back to zero, and then does the same thing in the other direction during one cycle.

A moving coil meter (as 99% of multimeters are) indicates what is called the 'rms' value of the voltage. This value (which stands for root mean square) is more or less an average, taking into account the fact that the voltage is always changing.

The 'rms' voltage has exactly the same 'work value' as a DC voltage of the same magnitude. In other words, if you supplied an electric heater with 240 volts rms, and then 240 volts dc, the heater would give out the same heat in both cases.

If you want to convert the 'rms' value your meter reads, to the 'peak' value (the maximum

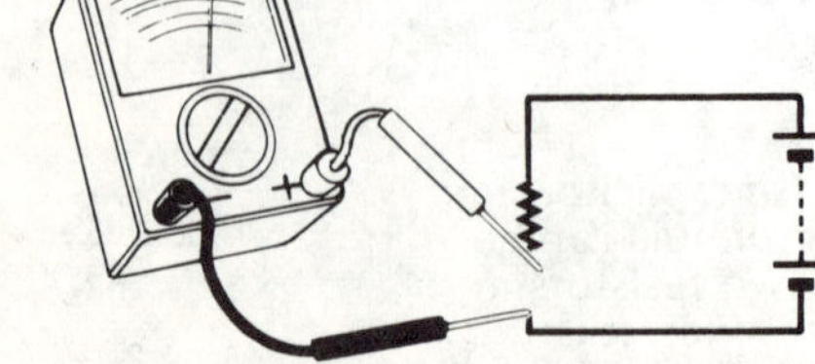
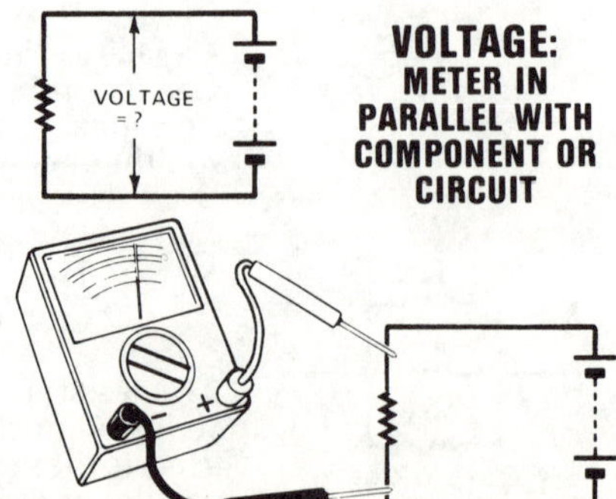

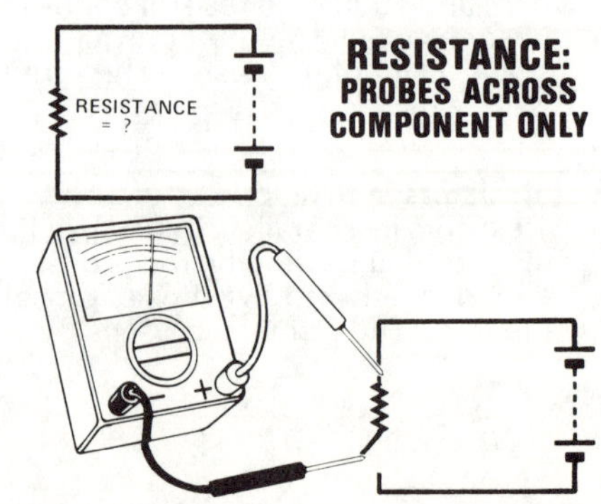

voltage reached in the cycle) you have to multiply by 1.4142. Conversly, the rms voltage is 0.7071 the peak voltage.

Measuring resistance:

When measuring resistance, it is important that the component you are reading is not affected by other components in the circuit. There is no point in measuring a resistor when there is another resistor in parallel which is interfering with the reading.

Therefore it is normal practice to remove one end of the component under test from the circuit to avoid any possible influence. Another wise move, even if you disconnect one end of the component, is to make sure power to the circuit is turned off.

Before commencing measurement (and each time you change resistance ranges) you should 'zero' the meter. This involves adjusting the knob on the meter so that the pointer reads exactly zero ohms when the probes are shorted together.

Select the lowest resistance range and zero the meter. Then place your test probes across the resistor (or other component you are testing). If the pointer doesn't move (or barely moves) switch up to the next range, and the next, until the pointer is in the last third of the scale. This is the area where best reading accuracy is obtained.

If the pointer reaches, or over-shoots zero, back off one range. Re-zero the meter and then read the value from the 'ohms' scale. Multiply the reading from the scale to the multiplier indicated by the knob: for example, if you read 15 on the scale, and the knob points to 'x100', the actual resistance is 15 x 100 or 1.5k

Checking transistors and diodes

While not a perfect check, a multimeter can usually give you a 'go/no go' test on most diodes and small signal transistors. (Testing power transistors is not quite so simple, as the results can sometimes be misleading).

Testing diodes: This is very easy. A diode should conduct in one direction only: if you set your multimeter to a low ohms range (say x10) and place your probes one on each end of the diode one way, the meter should read fairly low. Reverse the probes, and the meter should read fairly high. If both these checks are ok, then basically so is the diode!

Zener diodes can also be checked the same way. You can't tell their voltage, but you will at least have a go/no go indication.

Testing transistors Because transistors are basically two diodes in one package, we can check them in a similar way to diodes. We can also check for 'punch through' from collector to emitter.

Switch to a low ohms range, (eg x10) and connect the probes to the emitter and collector. Read the meter, and reverse the probes and note the reading again. Both should be high. If so, continue. If not, you should probably throw the transistor away!

Switch to a high (x1000) range, and connect one of the test probes to the transistor's base. Connect the other probe to the emitter, then collector, noting the readings. Both readings should be roughly the same; high or low, it doesn't matter (but remember which!)

Now swap the probe on the base with the other, and do the same check. The readings should still be roughly the same, but opposite to the last check. If the last readings were both high, these should be both low, and vice versa.

If these tests are ok, the transistor is also probably ok. By the way, you can work out whether your transistor is an NPN or PNP with these checks: with the black lead on the base and both readings low, the transistor is an NPN type. With the red lead on the base and both readings low, the transistor is a PNP type.

POINTS TO REMEMBER:

- **Meter in series for current**
- **Meter in parallel for voltage**
- **Meter across component only when measuring resistance**
- **Start on the highest range and work down**
- **Zero the pointer when changing to any resistance range**
- **Replace the batteries if you cannot zero pointer**
- **Use the mirrored scale to avoid parallax errors**
- **Swap the probes if the pointer swings backwards**
- **Take care of your meter!**

what's inside a multimeter?

As you can see from above, the multimeter is very versatile. But do you know how it performs all those functions?

The heart of the multimeter is the meter movement itself. This consists of a coil of wire, suspended on an axle, in a magnetic field from a permanent magnet. If a current passes though the coil, another magnetic field is set up. The two fields repel each other (just like two magnets can repel each other), and the coil tries to move.

The only way the coil can move is to rotate on its axle. Attached to the axle is a spring, and a pointer which can move over a scale. Because of the spring, the amount of rotation of the coil will vary precisely with the amount of current flowing through the coil. So the scale can be graduated directly in units of current.

The current though the coil is, however, very, very small. In a modern multimeter, it takes just 50uA to make the pointer travel from one end of the scale to the other. (This is called full scale deflection, or fsd).

Obviously, a multimeter can measure a lot more than 50uA (many meters measure up to 10A — 200,000 times as much!)

The basic idea used to let us measure higher currents is quite simple. If two resistors are connected in parallel, the current divides up in inverse proportion to their resistance.

Exactly the same thing happens with a multimeter. The switch connects various resistors in parallel with the coil so that the current between the coil and resistor divides. An average multimeter has a coil resistance of 2000 ohms: if a 2000 ohm resistor was connected in parallel, the meter would read full scale when 100uA was flowing. If a 0.00005 ohm resistor was connected in parallel, the meter would read full scale with a current of 1 amp flowing! (Yes, resistors of this accuracy are required, and used, in multimeters!)

The resistors switched in parallel with a meter movement are called '**shunts**' — they shunt most of the current past the meter.

Now, how about voltage? As we mentioned, a meter measures current. But the coil has a certain resistance, so from Ohm's law we know that to make this current flow a certain voltage has to be applied to the coil. With a 2000 ohm movement, it takes just 0.1V to make 50uA flow (for fsd).

It follows that if we add additional resistors in series, it will take more and more voltage to maintain that same 50uA current.

If we added a 2000 ohm resistor in series, the voltage would divide across each 'resistance' so there was still 0.1 volts across the movement with 0.2 volts applied. So the meter would read full scale. If we connected a 1,998,000 ohm resistor in series, we could apply 100 volts to the meter, and it would still only read full scale.

Of course, the pointer doesn't have to swing all the way; a lesser voltage will give us a lesser swing. So we can calibrate the scale directly in volts, with the maximum voltage being what that series resistor will allow for exactly full scale deflection.

The resistor in series is called a '**multiplier**': it obviously multiplies the voltage required for full scale deflection.

Measuring resistance

A meter switched to the 'Ohms' scale is basically still a voltmeter; the difference being that the voltage which drives the movement comes from internal batteries rather than the circuit under test. (Note that a multimeter needs batteries **only** for resistance measurement — they can be removed completely for current and voltage!)

As the battery voltage remains constant, any resistance which is introduced in series with the meter will reduce the current flowing, thus reducing the reading. The scale is calibrated directly in Ohms, and must be multiplied by the figure indicated by the switch pointer. The switch selects different multipliers so that very wide ranges of resistance can be measured.

We mentioned before the 'zero ohms' adjustment. All this does is 'fine tune' the multiplier so that effects of battery voltage and component aging can be countered. If the 'zero ohms' control cannot bring the pointer to the zero mark the battery should be replaced.

Meter sensitivity

All meters 'load' the circuit being measured. By this we mean that the circuit or component is affected by the current taken from the circuit to drive the meter movement — small though it be. For minimum loading, the 'sensitivity' of the meter should be as high as possible.

Sensitivity is quoted in 'ohms per volt': in general, the higher the better. A modern multimeter would have a sensitivity of around 20,000 ohms per volt DC (more than enough for the average hobbyist use).

What this means is that on each DC voltage range, the circuit 'sees' the meter as a resistor, having a value of 20,000 times the full scale voltage of the range: on a 1 volt range it would be equivalent to a 20,000 ohm resistor; on a 100 volt range it would be equal to a 2,000,000 ohm resistor, over the entire range.

AC sensitivity is always lower than DC: 7 to 10 thousand ohms per volt is typical. This is adequate for the hobbyist.

Assembly hints and tips

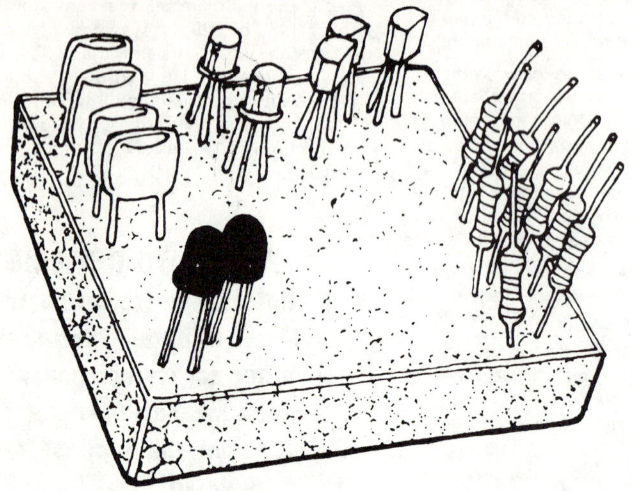

Figure 1: A block of styrene foam is the ideal way of holding components prior to insertion in the PC board. Except for CMOS IC's, of course!

As you gain experience in assembling electronic projects you will learn the best way to handle the components and how to achieve the best results most efficiently.

To save you a lot of trial and error and perhaps a few disappointments, we have set down a few hints that have been proven over the years to make assembly easier.

1. The first thing to do when constructing a project from a kit is to check that all parts have been supplied. Do this by ticking them off against the parts list shown with each project in this book.

2. Collect all the components in a dish or plate, or stick them into a block of styrene foam as shown in figure 1. This prevents them rolling off the table and getting lost or damaged.

3. Start construction by mounting the low profile components like links, resistors and capacitors first, taking care that they are properly 'dressed' or positioned to give your job a professional appearance. Good dressing simply means lining the components up so that wherever possible the codes are in the same direction and read from left to right—say, with the PCB facing towards you. When the PCB is turned clockwise through 90° all other components (normally running away from you) should again read left to right as the component layout suggests. Capacitors should read from the same direction as the resistors. And where components lay side by side, their coding should be oriented where they can be seen. Those components that are polarised must obviously be installed as indicated on the overlay or component layout diagram. Components should be straight and parallel. See figure 2.

Figure 2: This illustration of a computer board shows the strict method of 'dress' used – it enables rapid identification of circuit sections and components.

4. To achieve a professional result, care should be taken when installing components on to the PCB. See figure 3 for the correct installation of a resistor.

First measure the hole spacing and bend the component leads as shown with your long nose pliers. Then insert them through the top of the board and pull them through the copper side until the component rests flat against the top of the board. Bend the leads to slightly greater than 45° to hold the components on to the board, solder them and clip them off neatly against the solder. This allows them to be easily removed if necessary later on.

Another efficient method, and the more professional approach, is to load all low profile components like links, resistors and small diodes first. All leads are bent as described above then after all have been inserted, can be cut off as shown in figure 6. The board is then turned upside down and all the joints soldered. The higher profile components like capacitors and preset pots can now be inserted, cut and soldered. The last components usually inserted are the semiconductors like transistors and ICs. This is the most efficient way of inserting components but decide for yourself the method that suits your assembly procedure.

NB: Don't bend the component leads too much as they may suffer metal fatigue and break off. Also, don't bend the leads too close to

Figure 3 This resistor is partially inserted — the body will finally touch the top of the PC board and the leads on the copper side bent to retain the resistor in position prior to cutting.

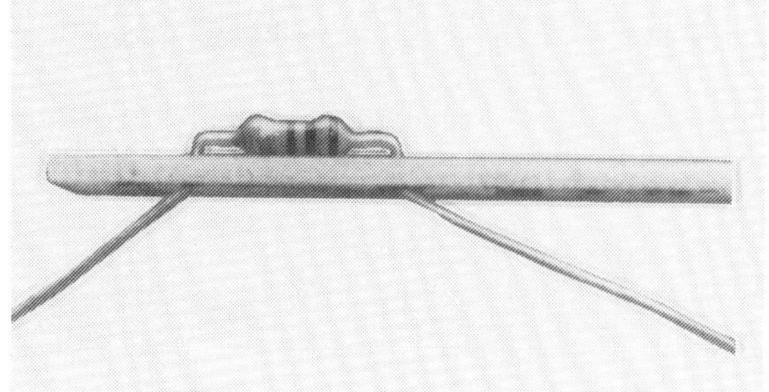

Figure 4. Component leads bent greater than 45 degrees prior cutting.

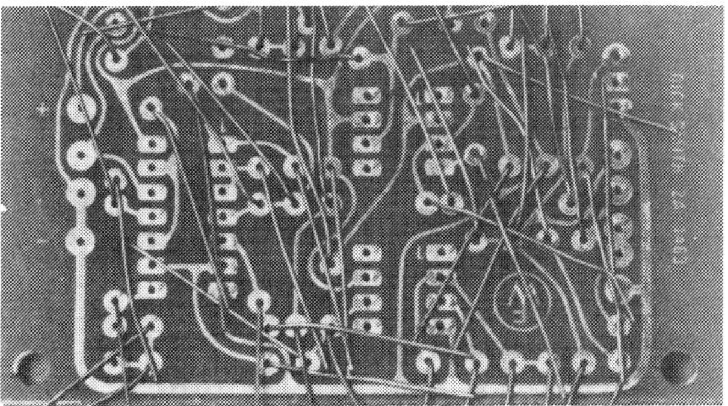

Figure 5. All low profile components inserted, leads bent, ready for cutting.

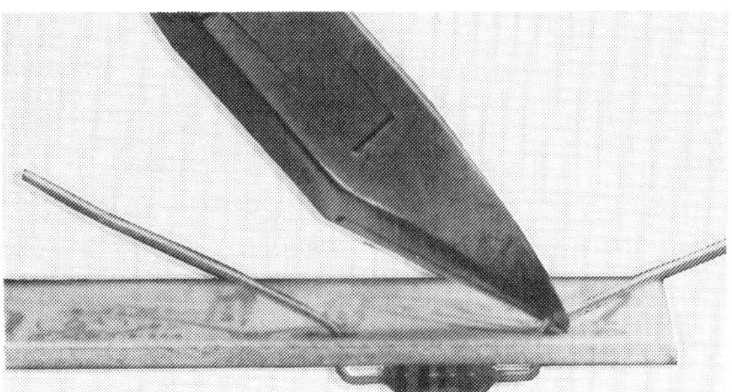

Figure 6 Lead is cut so that the remaining stub does not exceed the diameter of the PC pad.

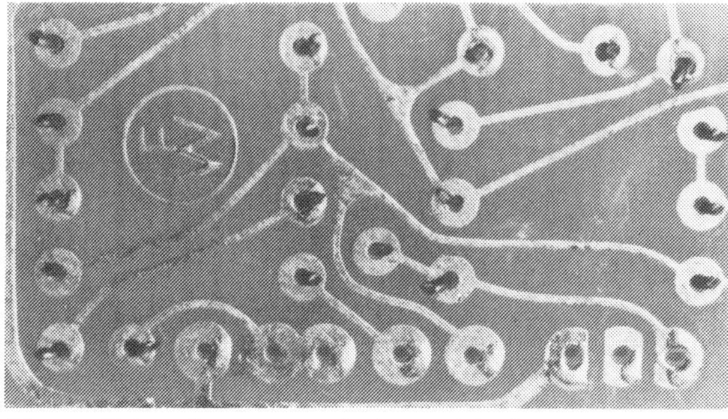

Figure 7 Leads cut, ready for soldering.

the transistor itself—especially transistor leads.

5. Be careful when cutting off a component's leads to make it fit better that you don't cut off the way the component's polarity is marked. This applies especially to Light Emitting Diodes (LEDs)—the negative (cathode or 'K') lead is normally shorter than the anode lead.

6. Offcuts of component leads may be used as wire links, so it is a good idea to save them.

7. It is not absolutely essential that components have the exact value shown in the circuit diagram. Both circuits and components are designed with certain tolerances and will probably work satisfactorily provided components are 'close enough'. Example: an electrolytic capacitor may be designated as 33 uF. In this case, a 30 uF would work perfectly. This applies even to transistors in most cases, as there are often several equivalent substitutes for any one transistor.

8. Before you connect your battery, double check all your components and wiring. Ensure that you have the correct components in the correct place with the right polarity, and that there are no other obvious faults.

9. You'll find it useful to start a spare parts box, to store bits and pieces that may come in handy when building up circuits in the future. You'll need to be fairly ruthless about throwing junk out though, or you could finish up with piles of useless bits and pieces.

10. When soldering to terminals, such as on speakers, the joint should be made mechanically robust first, so that the strain of the joint is taken by the wire itself, not by the solder. The solder merely forms a good electrical bond between the wire and the terminal.

Learning to solder

Learning how to solder correctly from the start will save you many hours of frustrating fault finding, as poor solder joints are the main reason that projects do not work.

Good soldering is the hallmark of the skilled electronics hobbyist and gives a sense of satisfaction in itself as well as going a long way towards ensuring that your project works correctly the first time you try it.

Good soldering requires the right iron properly prepared, the right solder, a correctly prepared joint and practice, practice, practice.

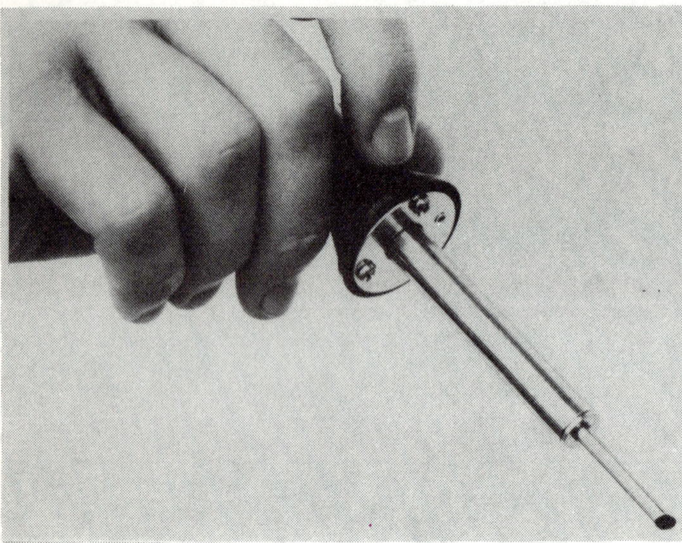

Figure 1. *The most important tool for the electronic hobbyist is the soldering iron. This illustration shows a typical hobbyist iron and the correct way to hold it.*

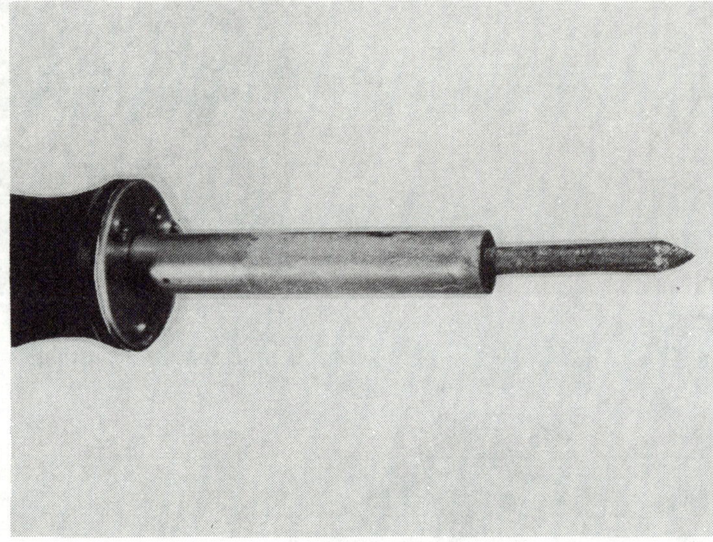

Figure 1a *It is important to keep your soldering iron clean at all times. This illustration shows a dirty, corroded bit that would certainly not aid you in obtaining a clean, secure joint.*

The Iron

An electric soldering iron of 10-30 watts rating with a 1.5 – 4mm chisel bit is ideal for the projects in this book. (See Fig.1). Such an iron should give years of service if reasonable regular maintenance is performed.

As most modern irons have plated tips it is not necessary (or advisable) to file the copper tip. If your iron has a plated tip simply switch the iron on and let it heat up. Once it has reached its operating temperature simply wipe the tip on a piece of damp cloth. This should expose a shiny surface. Apply a small amount of solder to this tip. Your iron is now ready for service.

Keep the tip clean at all times by periodically wiping it on a damp cloth or soldering sponge.

The Solder

The best solder for general electronics work is 60/40 solder (see 'components'). We recommend that you use a 60/40 solder that is pre-fluxed i.e. it has a small thread of flux running down the centre of the solder wire.

DO NOT use acid core solder or bar solder with Muriatic acid or 'Killed Spirits'. These fluxes and solders are commonly used by plumbers and tradesmen. Because of the corrosive nature of such fluxes they are not suitable for delicate electronic work.

Preparation

They say that there are three rules to be observed when soldering. They are: (i) Cleanliness; (ii) Cleanliness and (iii) Cleanliness.

Specifically though all parts of the joint to be soldered must be clean and free from tarnish, insulation and lacquers. If necessary use sandpaper or a fine file to clean the parts to be soldered. Most of your soldering will be components to Printed Circuit Boards (PCB's) or of wires to tags or lugs and these are usually tinned and clean so no surface preparation is necessary. This is the case with your Fun Way 3 Kits purchased from Dick Smith Electronics.

Component placement

The components are inserted as described in the previous section so that the resistors and capacitors are flush with the board and values are easily read (except for electrolytic capacitors where polarity must be observed).

Mount transistors proud of the board, while Integrated Circuits (IC's) are pushed in until the shoulders of the pins are flush with the top of the PCB. (See fig. 3). When you have a batch of components mounted you can turn the PCB over and solder the leads as described below.

Soldering

Holding the heated iron as you would a pencil (see fig. 1), place the tinned tip on the junction of the lead and the PCB track. Allow about one second for pre-heating and then apply the solder to the point where the bit contacts the lead and the PCB track as in figure 6. When just enough solder has flowed to cover the joint, slowly remove the solder and the iron together in a smooth action to leave a nice clean joint.

This is the most critical step and requires a lot of practice to acquire a high level of skill in soldering correctly.

The soldering of components to a PCB, tag strip or tracking board etc, will become much easier as you gain experience and confidence. For problem joints turn to the next page where the causes and the remedies are explained.

Note: When soldering semiconductor devices and other heat sensitive components spend as little time on the joint as possible. If necessary hold the lead with a pair of pliers or a heatsink clamp (see fig. 7), to prevent damage from overheating. (This will become unnecessary once you can solder quickly and reliably). Don't have the tip on the joint too long; overheating will destroy the flux and may damage the components or even the PC board. If the solder doesn't flow cleanly almost immediately the iron may be too cold or dirty.
An iron tip that is too hot will burn the flux and be indicated by large amounts of white smoke coming from the flux. Without flux the solder will not flow and therefore a poor joint will result.

Wiring to a tag

Ensure that the wire is clean and free of insulation and that the tag or lug is clean and untarnished.
Thread the wire through the tag until the insulation is up to it and bend the excess to make a good tight mechanical connection. Snip off the excess and the termination is now ready for soldering as described on this page.

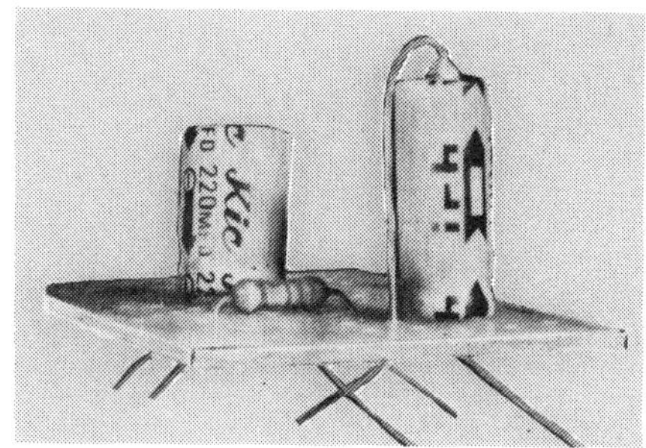

Figure 2   Components loaded prior to cutting.

Figure 3   IC's inserted.

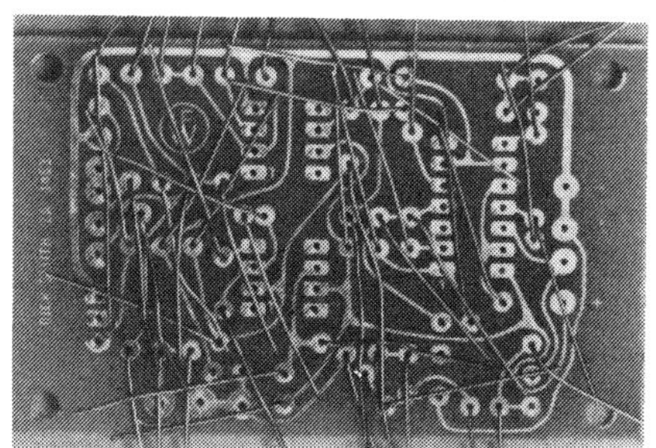

Figure 4   Leads ready for cutting.

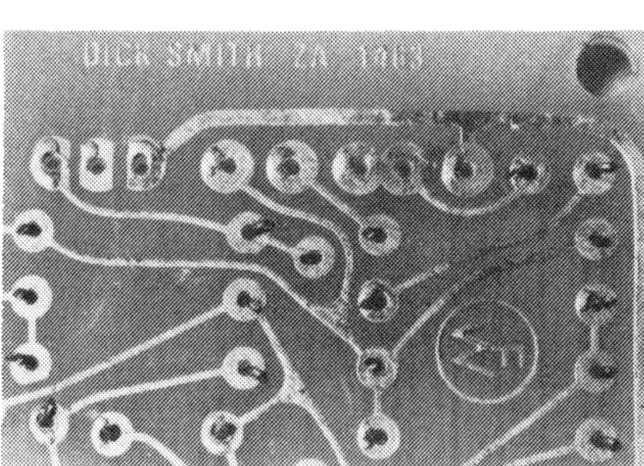

Figure 5   All component leads cut at edge of pad.

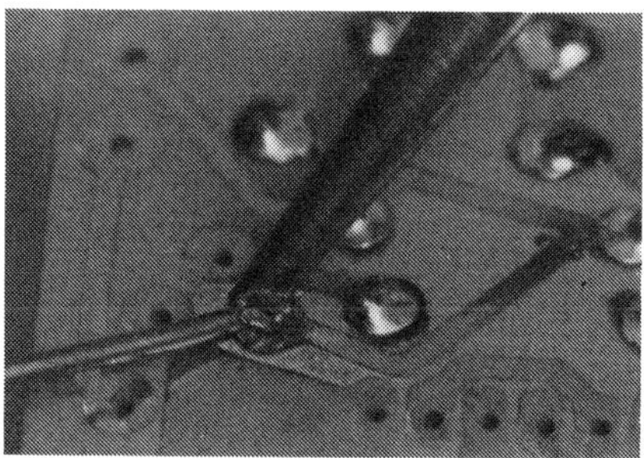

Figure 6   Soldering

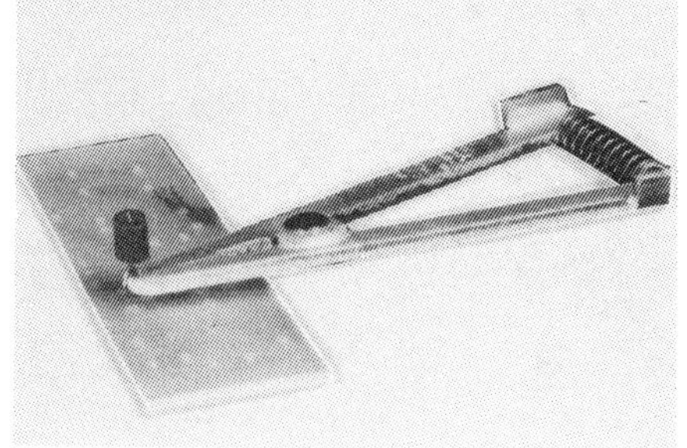

Figure 7   Some sensitive components may need a heatsink

Common faults, causes and remedies

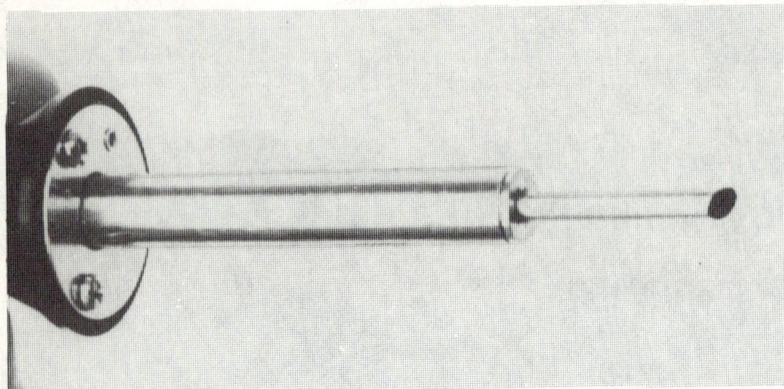

A clean iron tip is essential at all times, a damp sponge can be used to wipe the tip after soldering operations.

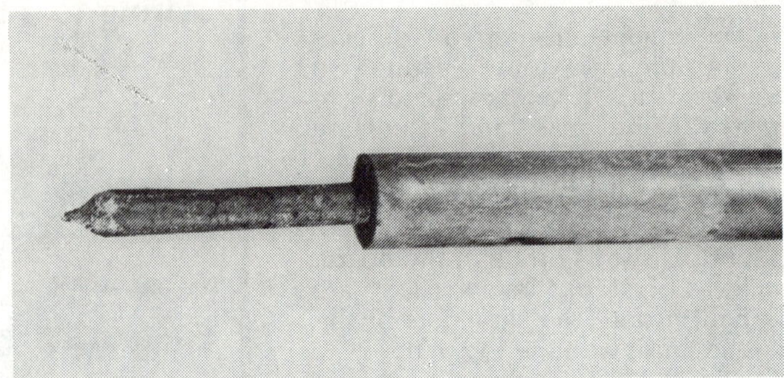

A dirty, badly shaped soldering iron tip can lead to bad solder joints. Reshape and clean tip.

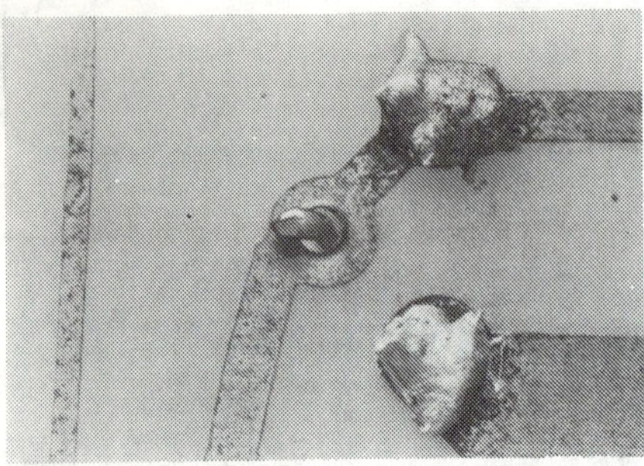

Corroded non-plated PC boards and joints that are made with little flux can end up like this. It is important that the components being soldered are clean. Flux must be present at all soldering operations.

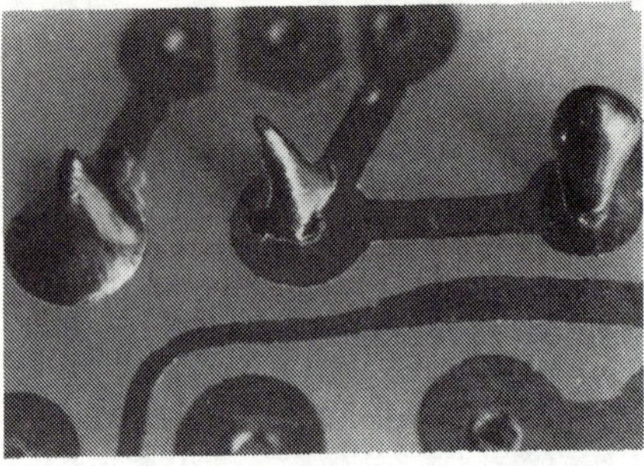

The two joints shown here are typical of a case where the solder tip and components being soldered are not brought together at the same time. It is also apparent that not enough solder has been applied.

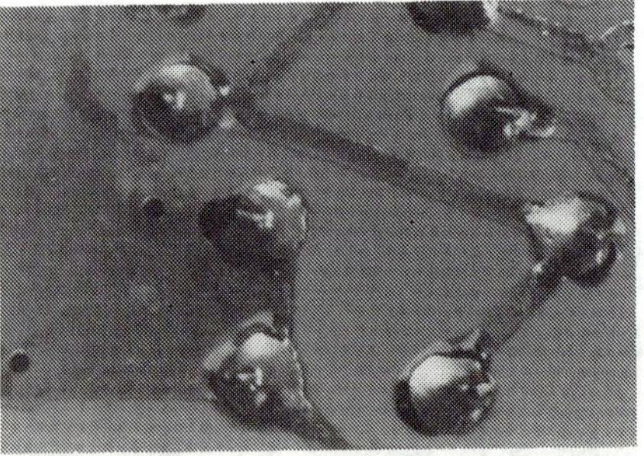

These finished joints are typical of a correct soldering operation. Note the shine on the body of the solder. The outline of the component lead is also visible in the smooth flow line. This is how all your soldered joints should look.

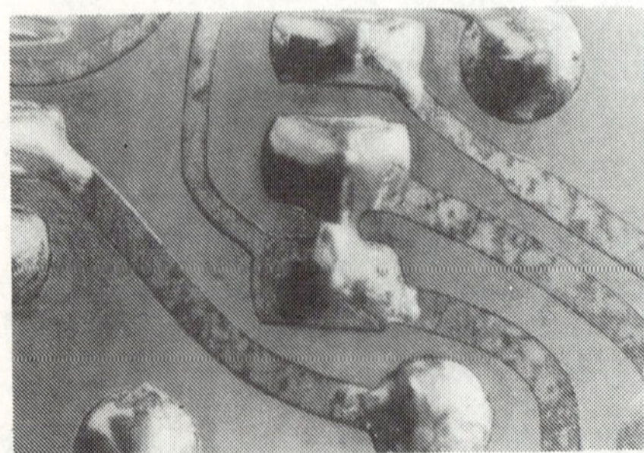

A case where the tip of the iron has been applied at the wrong angle with respect to the joint. The solder has bridged the adjoining pad.

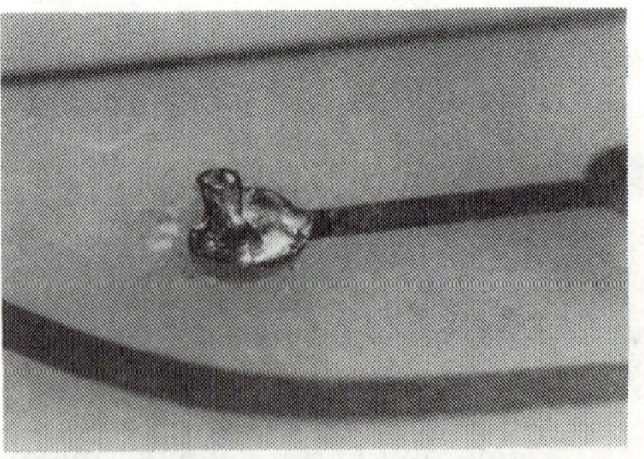

Physical damage to the PC pad is visible here. Too much heat, little flux and excess physical stress to the joint causes this problem.

Controlling other circuits with relays

A relay has several basic parameters which determine its usefulness in a circuit.

(1) Its coil resistance – this determines the range of voltage over which the relay wil reliably operate.

(2) Its contact rating – which determines the amount of current and voltage the contacts can safely switch.

(3) The number and type of contacts – some relays have just a single switching action which may be single throw or double throw (see 'switches' in the components section). Other relays may have a number of poles, with single or double throw action or even combinations of both. In most instances, the individual contacts of the relay are all insulated from one another, and can be used to switch different circuits without interfering with one another.

If, for example, you wish to switch two devices with one relay, you can do this with a 'double pole' relay (one having two sets of contacts). Relays are commonly available with up to four sets of contacts; special purpose relays have many times this number.

The S-7120 relay we use in the 'Fun Way' projects is a very simple type, chosen mainly because of its size: it fits nicely onto the printed circuit boards. However, it has only one single pole, double throw contact set (also known as a changeover contact or SPDT).

Let's say you want either more contacts or larger current carrying capacity, to use with one of the projects in this book. What do you do?

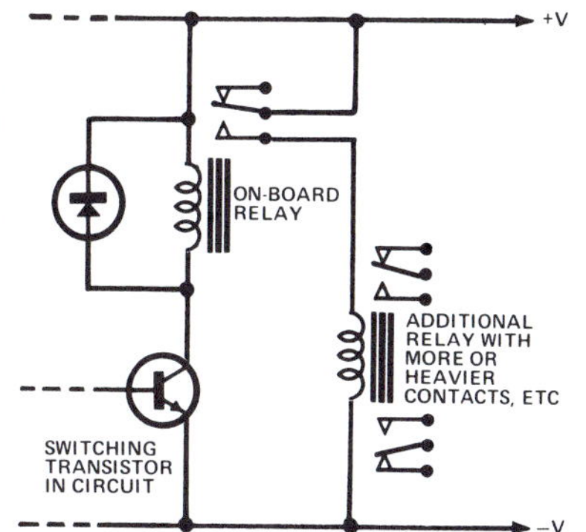

Figure 1

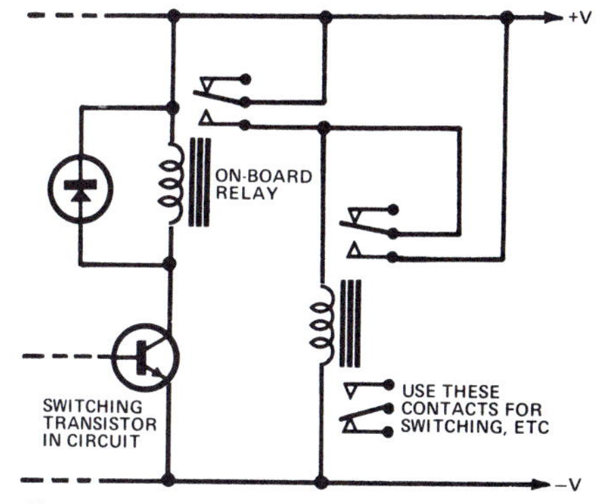

Figure 2

Of course, it would be possible to substitute a relay with the right coil resistance. All you need is one with a coil resistance of around 180 to 220 ohms, designed to 'pull in' at voltages from around 6 to 14.

Such relays are fairly common. However, size becomes a problem.

The easiest way for the beginner to overcome this problem is to use one relay to switch another: leave the one in the circuit as designed, and use its contacts to switch a second relay coil. This has a further advantage: if you want to 'invert' the operation of the circuit, you can use the 'normally closed' contacts and have the second relay operate in reverse to the way it would otherwise have done.

Figure 1 shows how to wire a second relay 'in tandem' with the one in the circuit. Because the second relay is being switched by another relay (not a transistor), it doesn't need the 'spike clamping' diode across it which you would normally use on a relay.

You can use any relay which will pull in with the voltage available from the battery or plug-pack adaptor (if you are using one). Just remember that a relay which pulls in reliably with a fresh battery may not be quite so keen when the battery gets a little flat!

You can use a relay with any number of contacts to suit the application. Ensure that you don't get solder dags across the contacts, as in some relays they are very close together.

Making a relay 'latch'. *(See figure 2).*

If a multi-contact relay is used, it is easy to make the relay latch on – that is, when it pulls in, it stays in regardless of what happens in the rest of the circuit. The only way to make the relay drop out again is to disconnect power.

This is done by using one set of contacts to bypass the switching transistor, thus keeping current flowing through the coil even though the transistor stops conducting. Once again, a diode is not necessary.

about the project kits . . .

Each of the projects in this book has an individual kit of parts made up for it: these are available from any Dick Smith store or re-seller. The catalogue number of the kit is given at the bottom of each component list.

To keep the cost of the kits to a minimum (and allow you as much flexibility as possible with your project) the kit contains only those components described in the main list.

In other words, most kits do not include Zippy boxes, batteries, speakers, sockets, Morse keys, or with the exception of a couple of kits, any on-off switches. Nor do they include any items of hardware, etc.

They do, however, include the printed circuit board, battery snap, (where applicable) plus all components required to make the basic project actually work.

Where potentiometers are specified, the kit may only include a pre-set trimmer type (even though the drawing might show a conventional shaft type). Once again, this is to keep the kit cost down.

Whether you replace the pre-set with a normal potentiometer, place the kit in a Zippy box, add on/off switches, external sockets, etc, is left completely up to you. Of course, should you decide to 'dress up' your project, all the components are readily available types,

and are available from your nearest Dick Smith Electronics store or re-seller.

The printed labels in the back of the book suit the commonly available Zippy boxes.

Please note that the kits do not include instructions: after all, you're reading this book and all the instructions are in it!

The contents of this book are copyright, and may not be photocopied or reprinted for any purpose without the written permission of the publisher.

How to make your own PC boards

What is a Printed Circuit Board?

Basically a PCB is a sheet of insulating material with a thin layer of copper laminated to the surface. The copper is etched away, where not required, to form the conductors of the circuit. The various components required in the circuit are usually mounted on the board on the opposite side to the copper, with their leads passing through small holes and soldered to the copper. Printed Circuit Boards are a far more efficient, reliable and compact method of connecting components together than previous methods of wiring.

The PCB's supplied in the kits that constitute the projects in this book, have been prepared professionally, but there is nothing to stop you preparing your own boards, either from the layout patterns in this book or from scratch using your own designs.

The un-etched board is as shown in figure 1, with a thin layer of copper attached to a sheet of insulating material, which may be of SRBP (synthetic resin-bonded paper) or fibreglass. More complex electronic circuits may require double-sided boards which, as the name suggests, have the copper on both sides of the insulating material. They may even have 'plated through holes', connecting the two copper sides together at various points.

A typical etched board is shown in figure 2, where the unwanted copper has been etched away by acid to leave the required circuit pattern.

PCB's vary considerably in both thickness of the insulating material and of the copper but the more common are around 1.5mm thick with 0.03mm - 0.07mm of copper.

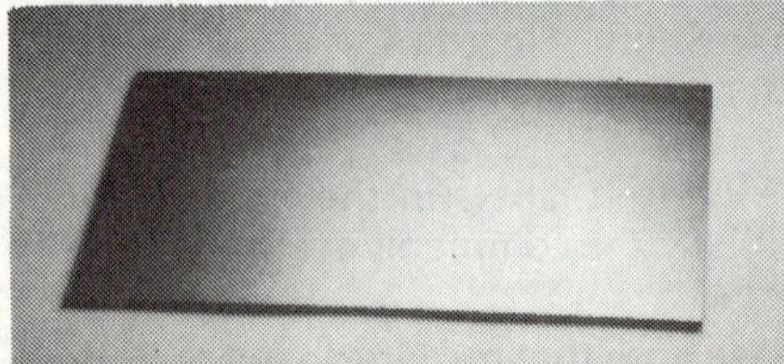

Figure 1

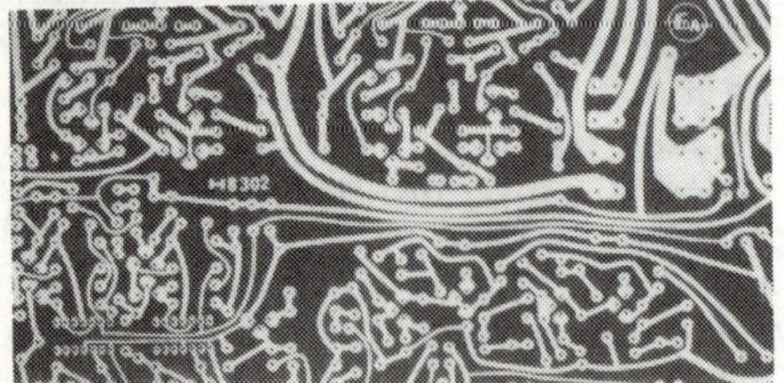

Figure 2

Cutting the un-etched board to size.

If you do not have a blank board of the required size you will need to cut a piece from a larger sheet. This may be done with a fret, coping or similar fine bladed saw, figure 3, or by scribing the board deeply along its dimensions on both sides and snapping off the required piece. The edge of a small screwdriver or the broken edge of a hacksaw blade make good scribers.

Using the edge of a steel rule as a guide, figure 4, scribe deep lines on both sides of the board directly opposite each other until the board is able to be snapped off cleanly. Dress the edges by filing to remove all burrs and roughness.

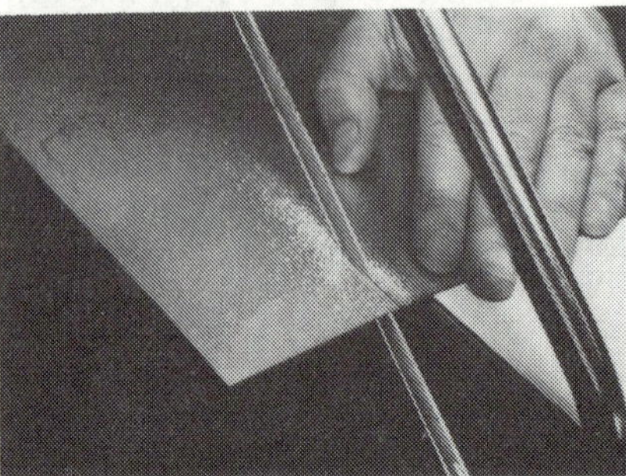

Figure 3

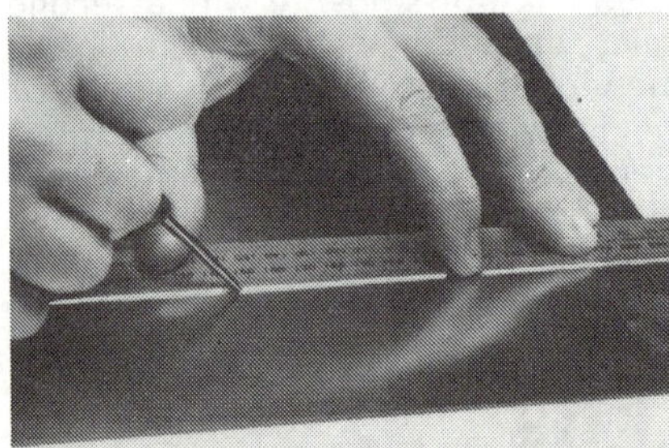

Figure 4

How to form conducting paths.

As we have said the un-etched board has copper completely covering the surface of the insulating base material, figure 1. To make the required 'tracks' and 'pads' the excess copper must be removed, leaving a pattern similar to the one shown in figure 2. This is done by coating the areas that are to remain with an acid resisting compound or material called a 'resist'. After this 'resist' is applied, the whole board is immersed in an acid solution known as an 'etchant', which will etch away any copper not covered by the 'resist' leaving only the required tracks and pads.

The PCB layout

Before the resist material can be applied to the board, a suitable layout must be worked out to join the circuit components together. This may mean using an existing PCB pattern, such as one of those provided in this book, or designing your own. Various methods are used but here is a simple method to start you off.

Using your circuit diagram as a reference (see figure 5) layout the components to give the easiest and simplest joining of the conduction tracks. You will find that using 0.1'' or metric graph paper as a layout guide will be a great aid as a left/right, up/down reference as well as providing stable reference for marks for component mounting holes etc. It is important to remember that your components will be on one side of the board while your tracks will be on the other (copper) side.

This layout stage is a matter of trial and error, but after a little practice it will become easier and you will find the results very rewarding.

Figure 5

Because this process is similar to printing photographs the circuits resulting are called 'printed circuits'.

When more than one board with the same circuit is required they may be produced photographically or by silk screening. This is the commercial method of production and is mentioned only as a matter of interest.

After the track layout has been done you can put in size reference marks to enable you to cut the finished board to the exact size you require. These marks are represented in figure 5 by the four crosses, one in each corner of

Making printed circuit boards

the board layout. It is only necessary now to mark on the tracks the positions of the component holes and this is done by making an oversize 'full-stop' on the track at the point where the components will be placed.

Cleaning the copper surface.

This simple but extremely important step is necessary to remove all dirt, corrosion, grease, etc from the surface of the copper so that the resist material will 'take' properly. One method is to apply a spirit such as metho, freon, alcohol, acetone or trichlorethylene to a clean rag and rub it briskly over the surface.

Another quite satisfactory method is to use a common household cleaning powder such as 'Ajax' applied with a clean rag soaked in warm water. As the powder acts as an abrasive it is advisable to rub in one direction only so that the fine scratches line up, giving the surface a brushed look. (See figure 6)

Once cleaning is complete flush the surface under a fast flowing warm water tap and immediately dry with a clean cloth.

IMPORTANT: From now on do not touch the surface of the board with your fingers or any dirty or greasy objects. Handle it by the edges as you would a record.

Figure 6

Preparing the board for the application of the resist material.

You are now ready to transfer your circuit layout to the copper in readiness for applying the resist material. As shown in figure 7 place a piece of carbon paper, carbon downwards, on to the copper foil so that it completely covers it. Place your prepared artwork over this so that the dimension marks line up exactly with the edges of the board. Fold the edges of the paper around the board and tape firmly in place with adhesive tape. Figure 8 shows the artwork being laid over the carbon paper prior to its being held down with adhesive tape.

Figure 7

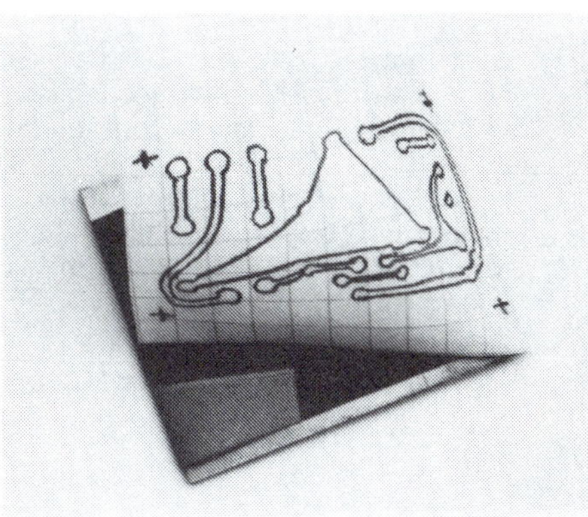

Figure 8

Now, with a pen or pencil, trace all the tracks and pads on the artwork. This will transfer the pattern via the carbon paper to the copper surface.

Remove your artwork (don't destroy it, you may need it) and carbon paper and the board is ready for the application of the resist material.

Application of resist material or compound.

Resist materials may take many forms but we will limit our discussion here to the three you are most likely to use: bituminous paint, Dalo pen and PC Board transfers.

Bituminous Paint

This may be applied with a number of brushes or pens, whichever you find easiest to use. As it is normally thick, you will need to thin it with small amounts of kerosene or mineral turpentine to a consistency where it will flow freely and evenly with the pen used. It may also be thinned by warming the container in a pan of hot water as shown in figure 9. This method will reduce the drying time of the applied solution.

Apply the paint evenly and smoothly, preferably after a few practice runs on a piece of scrap board and when you have followed all of the traced marks, allow about ½ an hour to dry.

Figure 9

Dalo Pen

As you can see from figure 10, the Dalo Pen is very similar to a normal felt tipped pen. It is a very convenient method of applying the resist solution, as you can virtually draw the circuit pattern on the copper foil. If the resist solution does not flow on smoothly, it indicates the surface is not clean.

To increase the flow of resist, push down on the tip to pump out more fluid. Allow 15 minutes to dry.

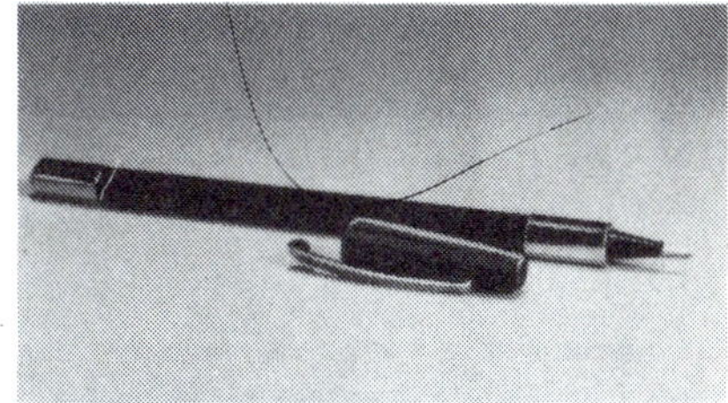

Figure 10

PC Board Transfers

These are pressure sensitive, rub-on tranfers applied to the surface of the copper foil with a smooth, round pointed instrument such as a thick ball point pen, knitting needle, or the end of a paint brush handle.

Take care that the transfer does not crack or peel as it is applied, as this will allow the etchant to penetrate. If cracking does occur, lightly scrape the transfer off and start again.

continued

Etching Materials

Various types of etchants are used commercially, including ferric chloride, ammonium persulphate, nitric acid etc. We will only discuss ferric chloride as it is the safest and most readily available. Nitric acid is not recommended as it is highly corrosive, difficult to store and very dangerous to handle.

Ferric Chloride

This may be obtained either as an anhydrous (dry) powder or a concentrated liquid. Both must be diluted by water to the required strength.

IMPORTANT: Follow the label instructions carefully when diluting either the solution or the powder. Add the powder to the water slowly, stirring as you do so. You will notice that heat is generated as the two come together. Approximate quantity mixing ratios are 500 grams of powder to each one litre of water.

Etching the Board

The etchant is poured carefully into a flat container of glass, porcelain or plastic, large enough to hold the board to be treated. Slip the board into the solution (see figure 11) and agitate from time to time, but careful not to breath in the fumes. After all traces of unwanted copper have gone, remove the board and wash it under a running tap before drying it off with a clean rag. Etching time will vary, depending on etchant strength, size of the board, amount of copper to be removed, temperature etc,. 10 to 30 minutes is common. Etching time may be reduced by agitating the etchant and heating the solution. (Not over 50 degrees). The solution may be used again, but it will become slower in action as successive boards are etched.

Figure 11

IMPORTANT: Etchants are corrosive and should be handled with care, with any spillage being cleaned up immediately. Used or waste etchant should be disposed of sensibly.

Drilling the Board

Once the board has been etched, washed and dried, you are ready to drill the holes. The drill point for the holes can be made with a centre punch. Place the centre punch so that the sharp point is in the position where the hole is to be drilled, and tap the head of the punch very lightly so that the indent made is just sufficient to penetrate the copper foil. An automatic centre punch with adjustable impact is ideal for this purpose (see figure 12).

Although the hole size requirements may vary because of component lead thickness, only a few drill sizes are generally required. The most common sizes are: No. 65 (0.83mm) for semiconductor devices e.g. IC's and transistors and No. 59 (1.0mm) for general components such as resistors and capacitors.

If you don't intend making a lot of PC Boards, a pin vyce may be used to hold the drill. The pin vyce is also very handy for enlarging or cleaning holes. (Vyce Cat T-5115 is ideal). For larger boards, or if you intend to make more boards, the handheld Minidrill (Cat T-4750/4751) unit is suited to this job.

Drill carefully, keeping the drill upright at 90 degrees to the board to make clean, straight holes. (See figure 13).

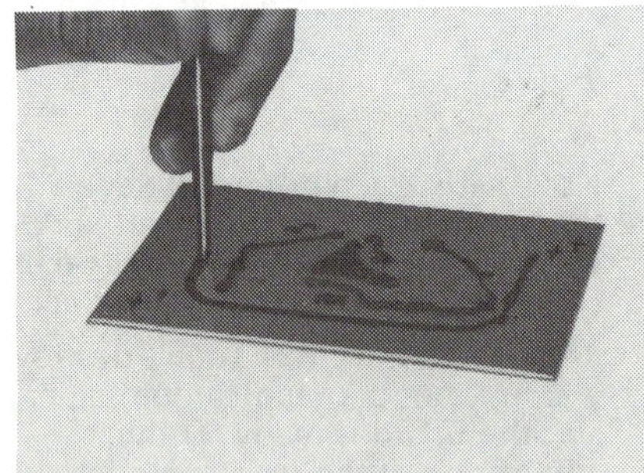

Figure 12

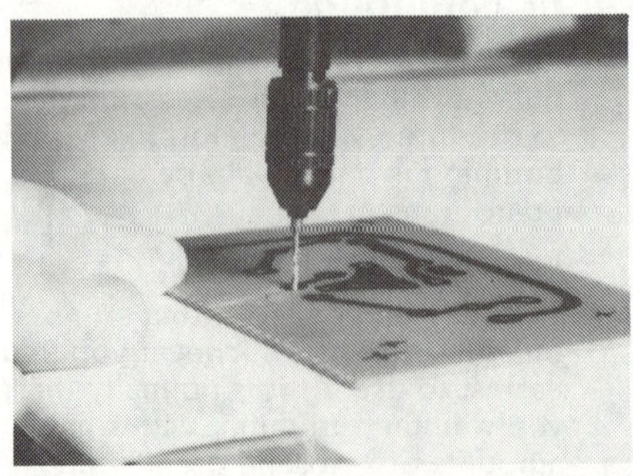

Figure 13

CAUTION: It is not advisable to use the normal hand held electric drill as they are too clumsy and broken drills may result. But if you have access to a high speed vertical drill press, this would be very suitable.

Final Cleaning

As in the original cleaning of the board, a household cleaning powder is suitable for this task. Rub only in one direction until all traces of resist and dirt have been removed (see figure 14) and then flush the board under a tap before drying with a clean cloth.

DON'T TOUCH THE COPPER SURFACE UNTIL IT HAS BEEN PROTECTED.

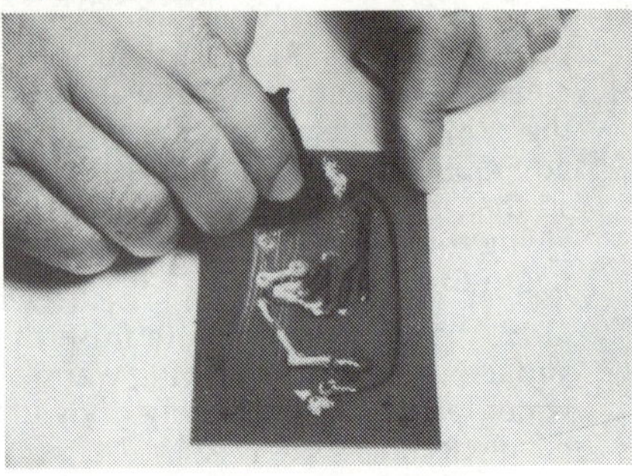

Figure 14

Surface Protection

To protect the copper surface against tarnishing during handling, a protective lacquer is sprayed lightly to form a thin film over the printed foil. (See figure 15). A suitable spray is PCL-2 circuit lacquer (Cat N-1045), or Clear Acrylic Spray (Cat N-1011). These sprays are solder through, which means that they will not hinder soldering on to the board.

Your printed circuit board is now complete and can be used immediately or stored for later use without deterioration.

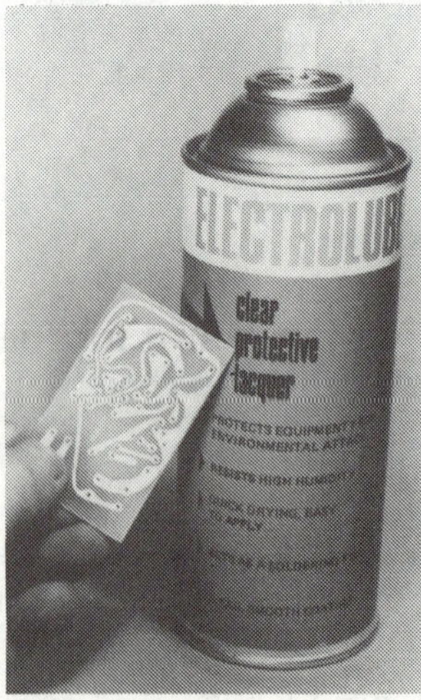

Figure 15

Projects

Light and Sound

Although simple in design, this versatile project can be adapted to suit many uses. The construction detail describes how it can be used as a LED flasher, mini strobe flasher, police siren and motor boat sound effects, continuity tester and Morse code practice oscillator. The circuit is based on one IC and the number of external components are kept to a minimum. The power supply can be as simple as two 1.5 V batteries and will happily keep the circuit going for days!

Right: Mini Strobe Flasher in optional Zippy Box

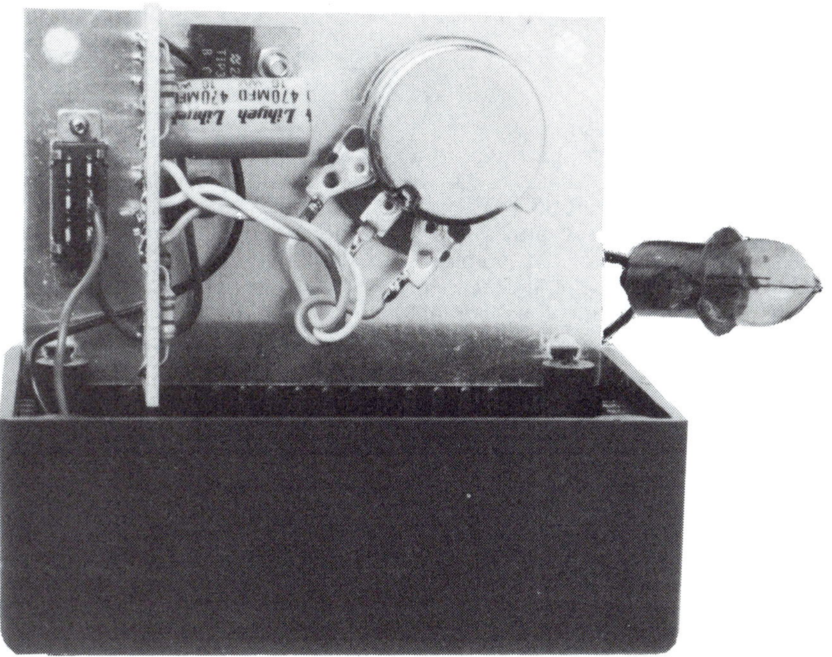

This photograph shows how the Mini Strobe Flasher is installed into the optional Zippy Box. Note the power transistor mounted on the front panel.

Each project is centred around an LM3909 integrated circuit that was originally intended as a timer/driver to allow operation of a LED (light emitting diode) from a 1.5 volt supply. Normally, the operating voltage of a LED is around 1.5 to 2.2 volts. The terminal voltage of a dry cell is 1.5 volts and therefore not capable of lighting such a semiconductor device.

The LM3909 changes this situation and at the same time opens up other areas of application. With only a few external components, a LED flasher can be assembled and operated on a single 1.5 volt cell. With simple circuit changes, the basic design can be used for other projects.

PARTS LIST

Part Description		Total Quantity
Resistors:		
68 ohm	R3	1
220 ohm	R1, R2, R4, R5, R6	5
5k	Linear Potentiometer VR1	1
Capacitors:		
2.2 uF	RB Vertical Electrolytic C3	1
33 uF	RB Vertical Electrolytic C4	1
470 uF	RB Vertical Electrolytic C1, C2	2
Basic Kit of Components: Cat K-2665		

Part Description		Total Quantity
Semiconductors:		
LM 3909	IC IC1	1
TIP32	PNP Transistor TR1	1
5mm	Red LED LD1	1
Miscellaneous:		
SPST	Pushbutton Switch SW1	1
8 ohm	Speaker	1
	Hookup wire 500mm	1
	PC Board ZA-1465	1

Parts Description	Total Quantity
Optional Components: (Not included in basic kit).	
Zippy Box Case UB5 H-2755	1
4.5V, 6.3V or 7.2V Pre-focus or MES Torch Lamp.	1
3mm/6BA Screw, Nut H-1684 pack	1
DPDT Miniature Slide Switch S-2010	1
Morse Key D-7105	2
Hookup Wire, assorted colours W-4010 Pack	1
Battery or Power Supply	1

HOW IT WORKS

Basic Integrated Circuit Operation

By way of explanation, the LM3909 equivalent internal circuit, together with the minimum external component count is shown in the diagram below.

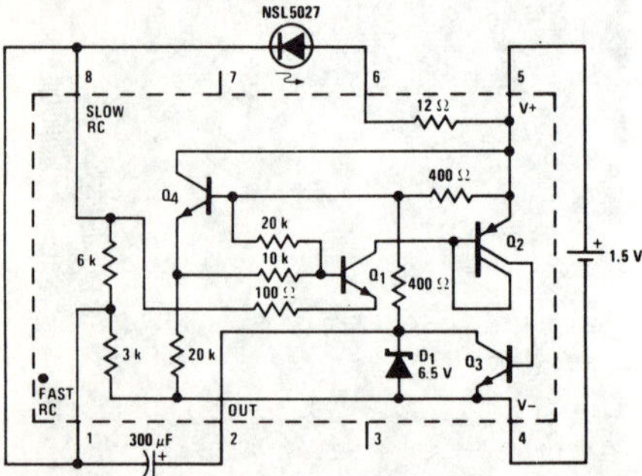

The flasher (National Circuit) achieves minimum power usage in two ways. Operated as above, the LED receives current only about 1% of the time. The rest of the time, all transistors except Q4 are off. The 20k resistor from the Q4 emitter to negative draws only about 50uA. The 300uf capacitor is charged through the two 400 ohm resistors connected to pin 5 and through the 3k resistor connected to pin 4 of the circuit.

Transistors Q1, Q2 and Q3 remain off till the capacitor becomes charged to about 1 volt. This voltage is determined by the junction drop of Q4, its base/emitter voltage divider and the junction drop of Q1. When voltage at pin 1 becomes a volt more negative than that at pin 5 (the positive terminal) Q1 begins to conduct. This then turns on Q2 and Q3.

The LM3909 then supplies a high current pulse to the LED. Current amplification of Q2 and Q3 is between 200 and 1000. Q3 can handle up to 100mA and rapidly pulls pin 2 close to the common negative supply line, pin 4. Since the capacitor is charged, its other terminal at pin 1 goes below the supply negative. The voltage at the LED is then higher than the battery voltage, the 12 ohm resistor between pins 5 and 6 limits the LED current.

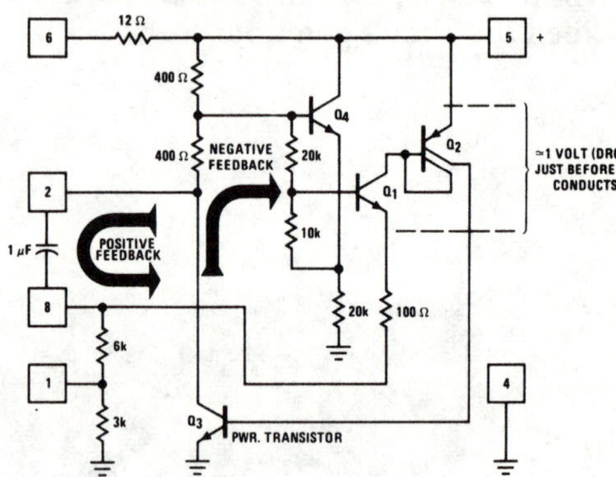

Many other oscillator circuits work in a similar manner. If voltage boost is not needed (with or without current limiting) loads can be hooked between pins 2 and 6 or pins 2 and 5.

GENERAL ASSEMBLY

The project you wish to build will dictate the parts that can be loaded into the PC board. The common parts to all projects you may wish to construct are IC1, R3, R4, R5, R6 and LD1. These can be inserted, pigtails cut and then soldered. See detail on each variation for other component details.

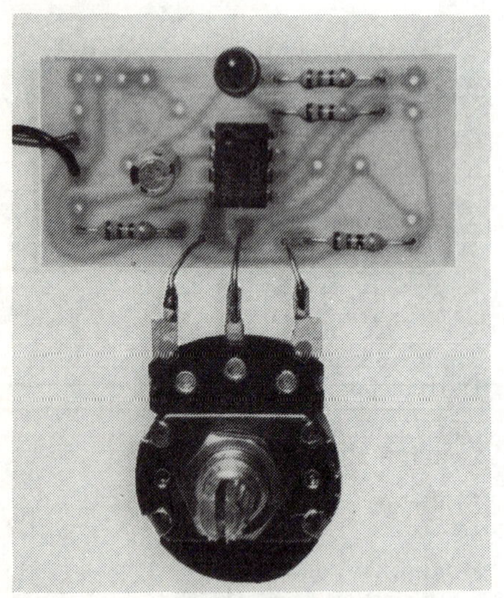

General Assembly of P.C. Board. The components shown are for sound effects.

VARIABLE RATE LED FLASHER FOR OPERATION ON 1.5 TO 6 VOLT

The most simple variation on the basic design is used in this LED flasher circuit. Insert C2, a 470uF electrolytic and the power wiring as shown on the PC board layout. This circuit detail is the basis of most of the projects described.

The timing/storage component C2, resistors R4, R5 and R6 together with the potentiometer VR1 make up the frequency determining section. The flash rate may be varied between 1Hz and 20Hz depending on the supply voltage. R3 is included in the circuit to limit the peak LED current to within safe limits for the diode and IC. It can be shorted out if the DC supply does not exceed 3 volts.

1.5 to 6 Volt Variable Flasher. P.C. Board Component Overlay Diagram

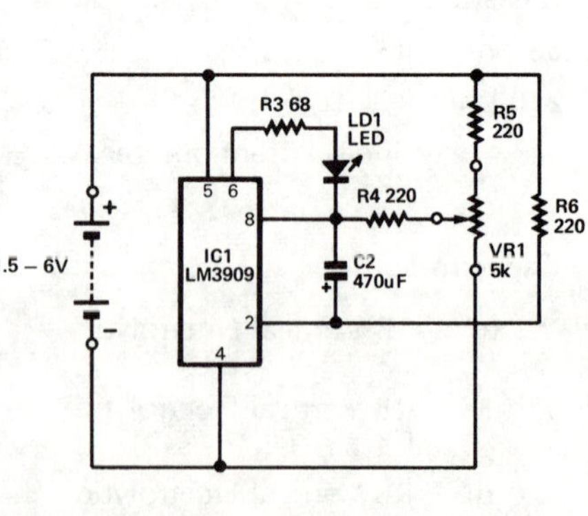

1.5 to 6 Volt Variable Flasher Circuit Diagram.

VARIABLE RATE LED FLASHER FOR OPERATION ON 6 TO 14 VOLTS.

This circuit has three extra components over that of the previous 1.5 to 6V design. As the maximum operating voltage to the LM3909 in this flasher mode is 6V, a divider made up from R1 and R2 is incorporated.

LED Flasher Circuit Diagram

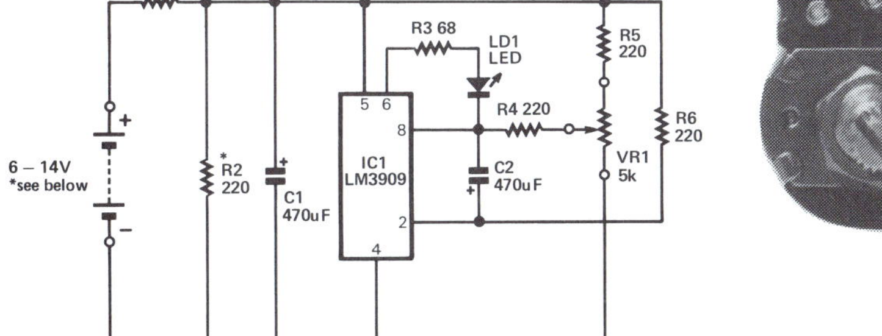

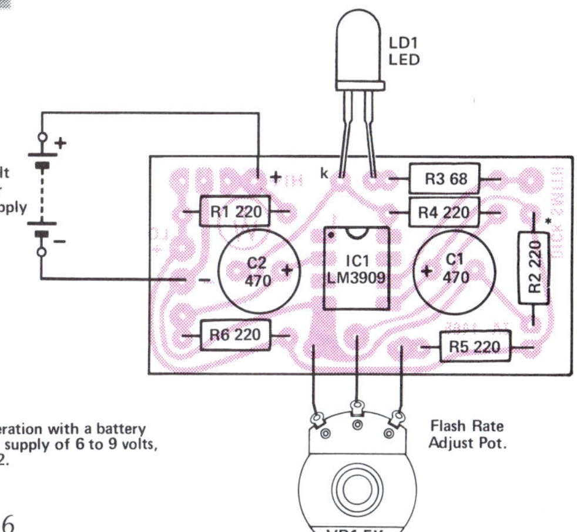

Photograph above shows completed 6 to 14 Volt Flasher Circuit Board.

For operation on supply voltages between 6 and 9V, R2 can be removed. The electrolytic capacitor C1 is used as a storage component to support the increase in current at the point of flash.

LED Flasher P.C. Board Component Overlay Diagram.

MINI STROBE VARIABLE FLASHER

This variation on the basic circuit uses a transistor to support the lamp current during the flash period. The lamp used may be any voltage between 4.5 and 7.2 volts depending on the supply used. These can be ordinary torch globes of the pre-focus or miniature screw type (not dial lamps) and are available at most stores where batteries and torches are sold.

R1 limits the circuit voltage to within safe limits of the IC and reduces TR1 base current. When the LED flashes, peak current flows in the supply lines and consequently a voltage drop occurs across R1. This voltage is also applied across the base/emitter junction of TR1 and as a result it is turned on. This lights the lamp and at

the same time effectively shunts most of the voltage available to the rest of the circuit over the period of discharge of C2. The voltage falls off at a rapid rate and returns to the original state to repeat the cycle.

Because of the high lamp current and the even higher in-rush current at the point of the lamp switching on, a power transistor has to be used in this application. A heatsink is required to dissipate excess heat from the transistor. If you decide to use the UB5 H-2755 Zippy Box, the aluminium lid is used as the heatsink in this case. The mounting hole for the transistor is on the left hand side at the centre (marked).

If you wish to use some other mounting or housing arrangement, an

equivalent surface area to this lid could be used as the heatsink. Not less than 1250 square millimetres should be used.

Constructors may wish to build this circuit into a 'lantern type' flashlight using an external switch to enable the flash operation. This could be the basis of an emergency flasher warning system. Again, some type of heatsink has to be provided.

As the flash rate can be varied up to around 20 flashes per second, the lamp could be used to strobe moving parts and effectively 'freeze' motion. This would be suitable for mechanical parts moving at a rate up to 1200RPM.

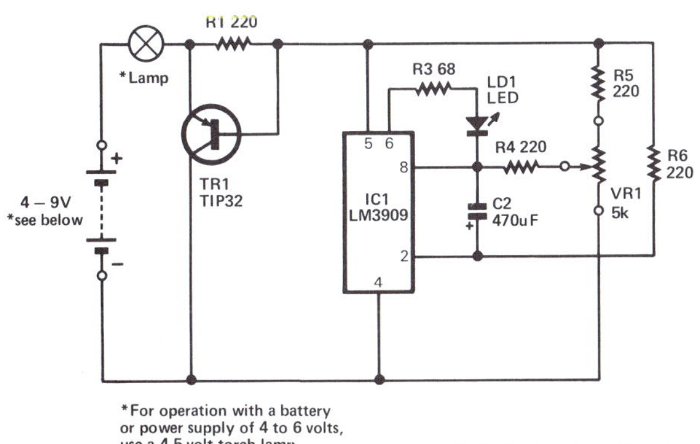

Mini Strobe Flasher Circuit Diagram.

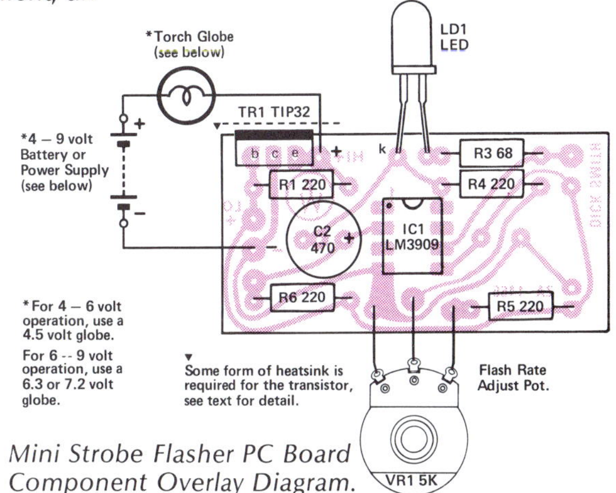

Mini Strobe Flasher PC Board Component Overlay Diagram.

SOUND EFFECTS

By the rearrangement and the addition of a few components, the basic circuit can be made to produce varying sound effects in a speaker or earphone.

By changing the storage capacitor 'C', different sounds can be generated. With a value of 33uF, the sound will roughly imitate that from an old fashion motor boat. A value of 2.2uF will change this sound to a fire siren.

The frequency rate adjustment potentiometer VR1 and the 470uF electrolytic capacitor, together with a pushbutton switch form the basis of the modification. The pushbutton is

Sound Effect P.C. Board Component Overlay Diagram.

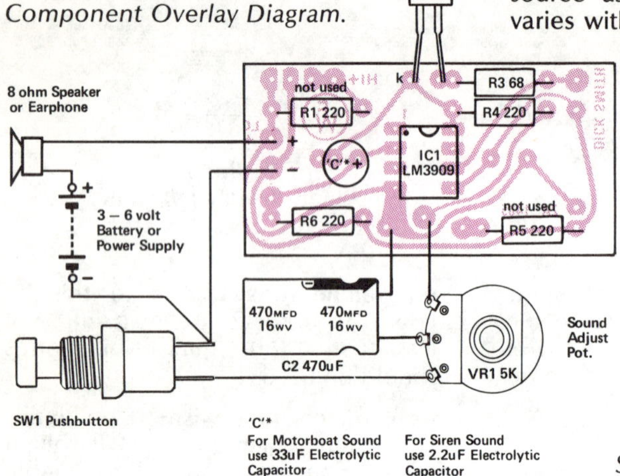

used to 'speed up' the sound while the 5k pot slightly varies the sound effect. The 8 ohm speaker or earphone in series with the positive supply lead becomes the sound source as the oscillation frequency varies with the current in this line.

Photograph (above) shows Sound Effects Assembly

This assembly can be housed in a UB5 H-2755 Zippy Box. The components can be mounted as indicated on the label for this project at the rear of this book. An optional on-off switch can be provided for convenience. The speaker would be external to the box, with 3.5mm speaker socket be used to couple it.

Sound Effects Circuit Diagram.

HOUSING ARRANGEMENTS

As we have suggested, most of the projects can be assembled in a UB5 H-2755 Zippy Box. The front panel label cut from the rear of this book can be used to locate the centres of components mounted on the lid. Place the label accurately over the lid and mark the points required with a sharp scriber or similar instrument. Only mark the points applicable to the project you wish to house. Now drill each hole as required, the LED is 5mm, the rate adustment potentiometer is 9mm, the pushbutton is 6.5mm to 7mm and the transistor mounting point is 3mm. If you are to incorporate an on-off switch, 2 x 2mm holes are required for this mount. The rectangular cutout will have to be made with a piercing saw or simply filed out with a small needle file.

After holes are drilled, the label can be glued to the lid with a rubber based adhesive. The holes can be punched in the label by using the rear shank of the drill to suit the hole.

The PC board is placed in the last slot position in the box. The legs of the LED will have to be bent at right angles to accommodate the body through the mounting hole. If the power transistor is part of the application, it will have to be located and soldered to fit over the hole provided for the mounting point. As the collector of the transistor is at ground or negative potential, no insulating material is required between it and the mounting surface. A 3mm/6BA screw, nut and washer can be used to fasten this assembly. The rate adjust pot will require short lengths of wire to connect to the circuit; a knob can be fitted if you wish. A hole will have to be drilled in the case to allow access for any external wiring.

This illustration shows how all components are mounted in the optional Zippy Box. Note that not all the components shown will be used on each project.

MORSE COMMUNICATOR

With slight modification and the addition of two Morse keys and an extra 8 ohm speaker or earphone, the circuit can be made to operate as a Morse trainer. The basic LED flasher circuit is used with a change of C2 to 2.2uF.

The local speaker is connected across pins 6 and 2 to the points shown. The local key and the supply battery are connected to the three wire line to the remote station. This station can be located up to 60 metres from the local unit. Virtually any wire could be used for this link.

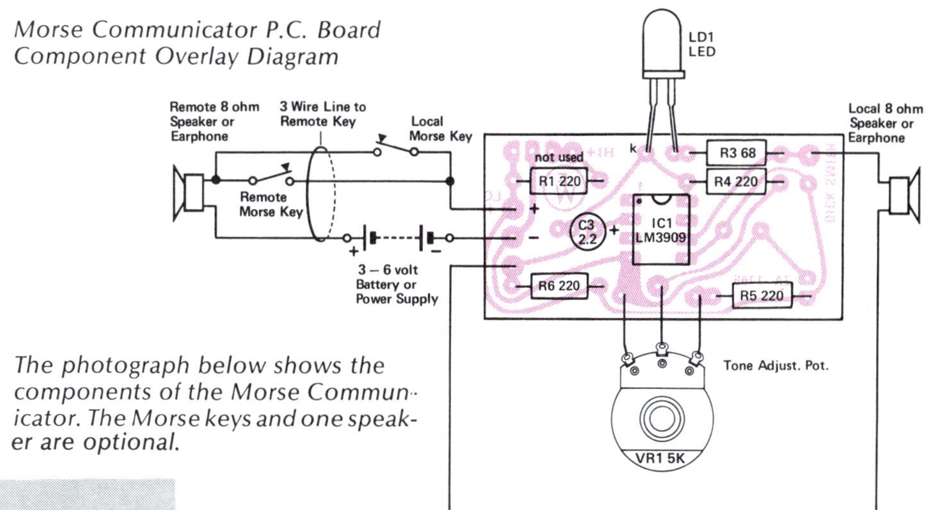

The photograph below shows the components of the Morse Communicator. The Morse keys and one speaker are optional.

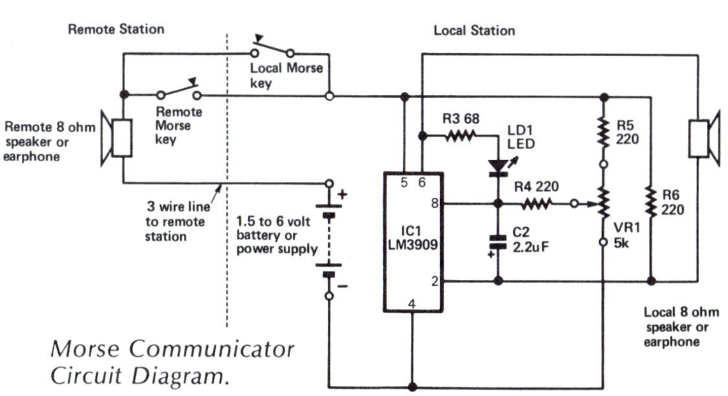

Morse Communicator Circuit Diagram.

CONTINUITY TESTER

Only a slight modification is required in this application to make a handy tester that will give audible indication of circuit continuity. The circuit can be operated on a supply of 1.5 to 6 volts and will give an output when the points being tested range from zero to tens of ohms in resistance. With a little experience in use, the difference in sound between ohms and tens of ohms resistance

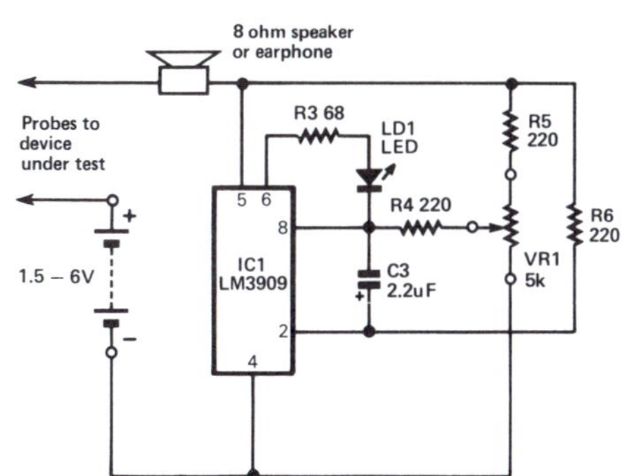

Again the UB5 H-2755 Zippy Box could be used to house the assembly. An AA size battery could be snuggly fitted inside the case if the supply leads are soldered to each end. It may require some insulation to stop it from 'shorting' against the board. For the sake of convenience, a 3.5mm speaker socket could be installed to connect the external speaker or earphone.

Left above; Continuity Tester Circuit Diagram.

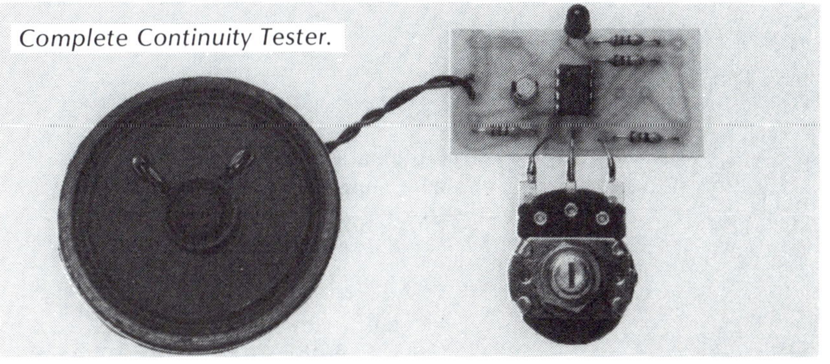
Complete Continuity Tester.

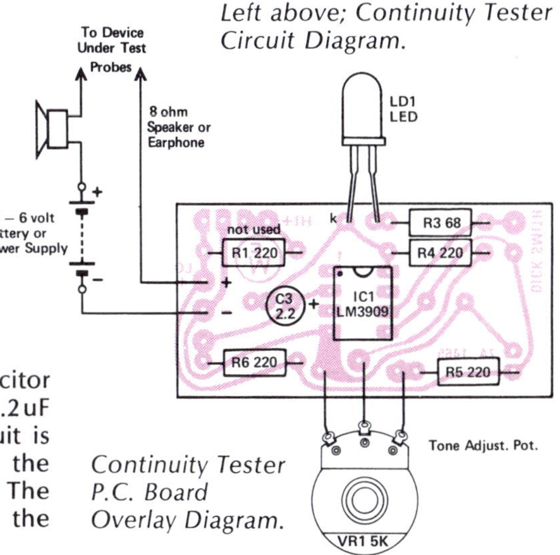

Continuity Tester P.C. Board Overlay Diagram.

measured will be recognized. This makes it a handy tool for work on automotive, electrical, intercom, line maintenance and similar service areas. Of course, all power to the equipment or wiring under test must be removed before proceeding.

The timing/storage capacitor component C3 is in this case a 2.2uF electrolytic. The rest of the circuit is left unchanged from that of the normal low voltage flasher. The speaker is wired in series with the supply lead as shown.

Two Up

The completed project shown in the optional Zippy Box

Despite the fact that playing two-up is illegal in most areas of Australia, in a nation of gamblers, this game has come to be recognised as the most distinctly Australian of all gambling.

There are, in fact two different versions of the game. The first, called 'swy' uses three coins so that there will always be a result – two out of three, or three out of three wins. The other version, the one we have modelled our game on, is the genuine 'two up' – where two coins are used.

In this version, three results are possible: a head spin (two heads), a tail spin (two tails), and a no spin (one of each). The spinner, in the centre of the ring, must always spin for heads. He (she?) bets a certain amount by placing the money on the floor in the ring. A 'tails' better is required to 'cover the centre' by placing an equal amount of money in the rings. Once the centre is covered, 'side bets' are allowed around the edges of the ring, with the side betters waggering amongst themselves for a head or a tail spin. Two pennies are placed on a special wooden throwing aid called a 'kip', normally with their heads facing up. (In fact, in real, live, genuine two-up games even the pennies are from a particular batch – King George pennies are used. (Not just any penny will do!)

When all bets are placed, the ringmaster calls 'Come in Spinner' and the spinner tosses the coins in the air, making sure that the coins clear the spinners head, and that the back of the spinners hand is seen as the coins leave the kip.

The coins must land inside the ring. If they don't, or one of the above rules has been broken in the toss, OR if the pennies land one head, one tail, the ringmaster calls 'No Throw' and the pennies are re-thrown. If the heads are thrown, the ringmaster calls 'heads' and the spinner can stay for the next throw. Bets are settled, and the 'house' takes a small percentage from the winners. If tails are thrown, the ringmaster calls 'tails' and the spinner is dismissed. Bets are settled, but the house doesn't take a percentage from the winners.

PARTS LIST

Parts Description		Total Quantity
Resistors:		
470 ohm	R1, R6, R10, R12	4
1k	R2, R7	2
10k	R3	1
100k	R4, R8	2
820k	R13	1
1 Meg	R5, R9, R11	3
Capacitors:		
.01uF	Ceramic C5	1
0.1uF	Ceramic C1, C7	2
.015uF	Greencap Polyester C3	1
0.47uF	Tag Tantalum C2	1
10uF	RB Vertical Electrolytic C4	1
100uF	RB Vertical Electrolytic C6	1

Parts Description		Total Quantity
Semiconductors:		
4009 CMOS	IC1	1
4011 CMOS	IC3	1
4013 CMOS	IC2	1
1N4148 Silicon Diode	D1	1
3mm Red LED	LD1, LD3	2
3mm Green LED	LD2, LD4	2
Miscellaneous:		
Mercury Switch	SW1	1
Miniature DPDT Slide Switch	SW2	1
Battery Snap	P-6216	1

Parts Description		Total Quantity
4 x AA Battery Carrier	P-6124	1
P.C. Board	ZA-1461	1
Optional Components: (Not supplied in basic kit).		
Zippy Box Case UB3	H-2753	1
AA Cells		4
Piece of foam packing approx. 55 x 35 x 35mm (to hold battery carrier if necessary)		1
Basic Kit of Components **Cat K-2661**		

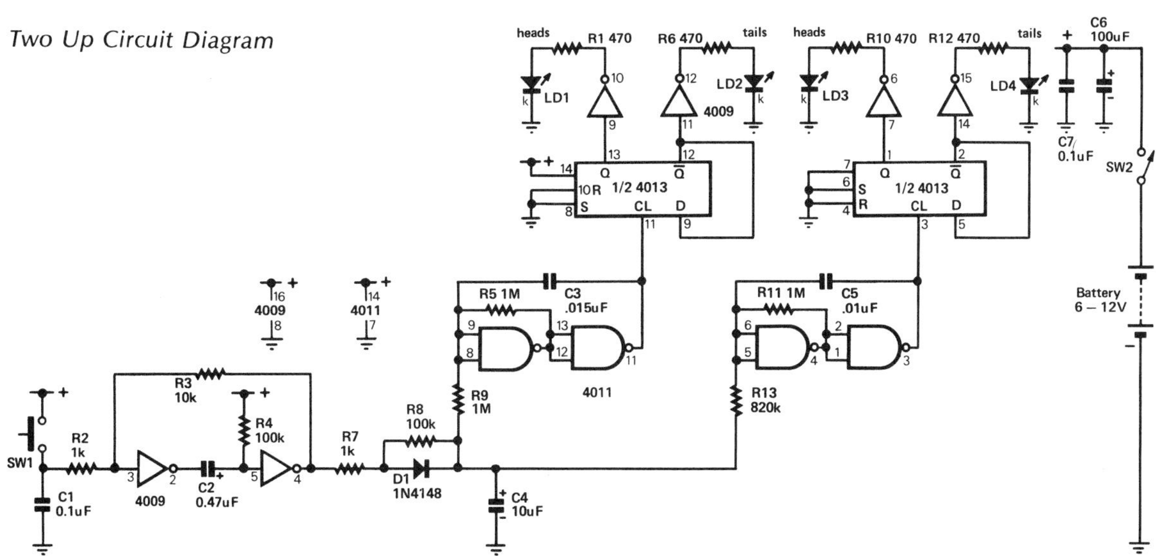

HOW IT WORKS

The electronics of this circuit has to simulate the toss, spin and final settling position of the two coins. The result can be one of three states, two heads, two tails or head and tail.

The face of the coin is represented by a LED (Light Emitting Diode), LD1 to LD4. The settling position of the coin is held in a temporary memory, in this case two flip-flops in a 4013 IC. The spin is simulated by two simple voltage controlled oscillators using a 4011 as the active device.

A mercury switch (SW 1) is employed as the toss mechanism, the height and time of the spin being controlled by a monostable made up of two inverters from a 4009.

In order for the contacts of the mercury switch to be closed, a simulated toss has to be made to get the mercury to 'make' with the contacts. This switch action applies the rail voltage (a '1') via R2 to the first inverter of the monostable. Normally the input of this inverter,

pin 3, is held low via R3 and the output of the second inverter. With a temporary '1' applied to pin 3, the output at pin 2 will go low and in turn, cause the input of the second inverter at pin 5 to follow via the timing capacitor, C2. The output at pin 4 will go high and remain in this state for a period set by the charging action of C2 and R4. The state of the mercury switch no longer has any control over the toss action.

This high state on pin 4 charges the 10uF capacitor, C4, via R7 and the forward conducting diode, D1. This voltage is stored in C4 and represents the time that the coins are in the air spinning.

The monostable will return to its initial state after a short "time out period."

The two sets of 4011 NAND gates configured as inverters (used because of the handy pack size) act as voltage controlled oscillators, the voltage being sourced from the

storage capacitor C4. As this voltage is discharged by R9, R13 and back through R8, the oscillators slow down and eventually stop. This simulates the slowing down of the spin to the point where they hit the ground. You may notice that the spin tends to "hic up" at some stage. This is when the two oscillators interact to further upset the final result. This may be seen as where the coins hit the ground and bounce.

The output of each oscillator is connector to the clock input of the flip-flops. At the end of the 'spin action', the last face of the coins is registered and held. The inverters connected to the outputs Q and Q- of the 'coin face memories' act as buffers to drive the LEDS, the 'face indicators'.

If you wish to increase or decrease the time that represents the period of spin, simply change the value of the storage capacitor C4. To increase the time, increase the value.

ASSEMBLY

1. Check your kit of components against the parts list. Values and quantities should match this list.

2. Mount the components as shown in the overlay diagram. All the resistors are inserted first. If you wish, they can be soldered after checking that all the values are in the correct position. Insert diode D1, observe polarity. Now insert all capacitors taking care with values, positions and polarity of electrolytics. Soldering can now be done.

3. ICS can now be put in (take care of the orientation). Remember that these devices are CMOS and should be handled with care. (See the section in other pages of this book on handling.) Solder into position.

4. The mercury switch SW1 is now wired. Be careful not to apply excess presure to the glass envelope, or to the leads as they are bent. A small piece of wire can be used to strap the envelope firmly in position. Solder in place.

5. Thread the battery leads connected to the snap through the hole in the PCB as shown. Solder to the board.

The ON-OFF switch is soldered to the rear of the PC board. It is simply a matter of using three of the previously cut pigtails from resistors to secure it in position. With your pliers, bend over the end of each wire so that a small 'L' is formed. Now insert each wire through the switch contact position from the component side of the board and solder into place. The switch is now threaded onto these wires so that the body lays against the PCB. Held in this position, the wires are soldered onto all lugs of the switch. Cut off excess pigtails.

6. The only components left to be fitted are the LEDS. Their physical position is dictated by the housing arrangement you wish to adopt. We have used a UB3, H-2753 Zippy Box for the assembly and have provided a front dress panel at the rear of this book to complete your project.

7. Case and Front Panel

Carefully cut out the label from the page and position it accurately over the Zippy Box lid. With a sharp instrument such as a scriber or centre punch, mark the centre positions of the LED holes and the

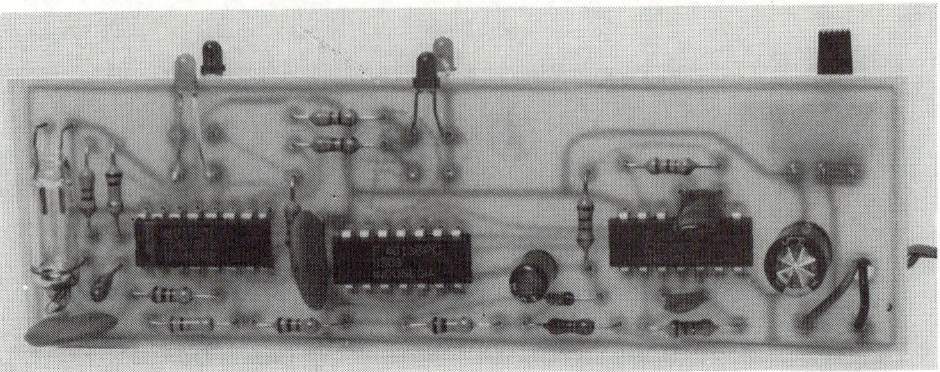

Complete Circuit Board

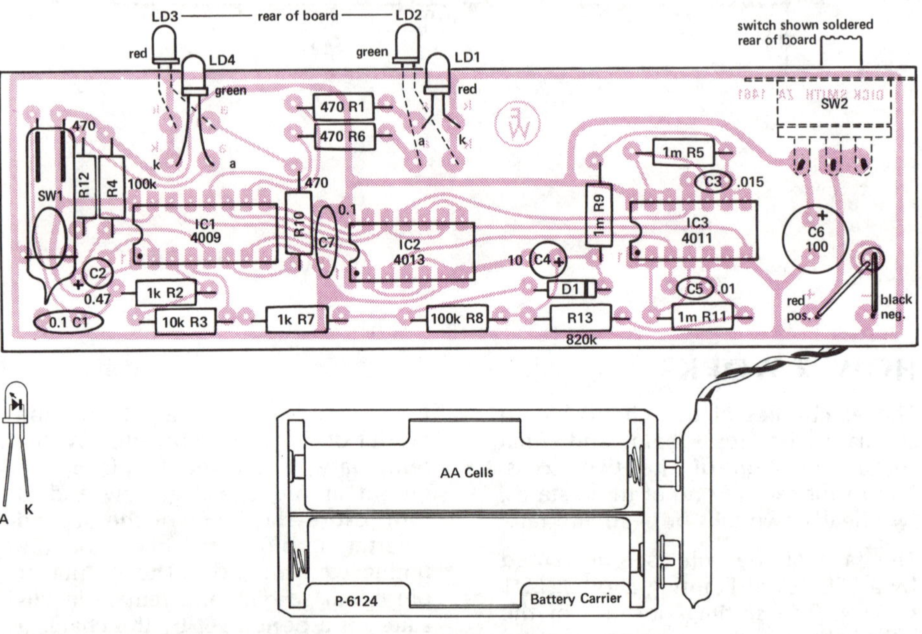

Printed Circuit Component Overlay Diagram
Note the LED's and switch on the copper foil side of the board.

corners of the ON-OFF switch hole, remove the label. Now drill 4 x 3mm holes for the LEDS where marked. Again, with a sharp instrument, join the 4 dots of the switch hole position to form a rectangle. Now drill a suitable small hole in the centre of this rectangle to accommodate a piecing, jewellers or fret saw blade. Carefully cut out the hole on the inside of the lines. The hole can now be dressed with a small needle file or similar. If you haven't a suitable saw, a small needle file could be used to shape this hole.

Now glue the dress panel to the lid with a suitable rubber based adhesive. Make sure that it is placed accurately in position. The holes for the LEDS and the lid mounting points can be punched out by pushing the back end of the previously used drill through each hole from the front face. A razor blade or similar knife can be used to carefully cut out the switch hole.

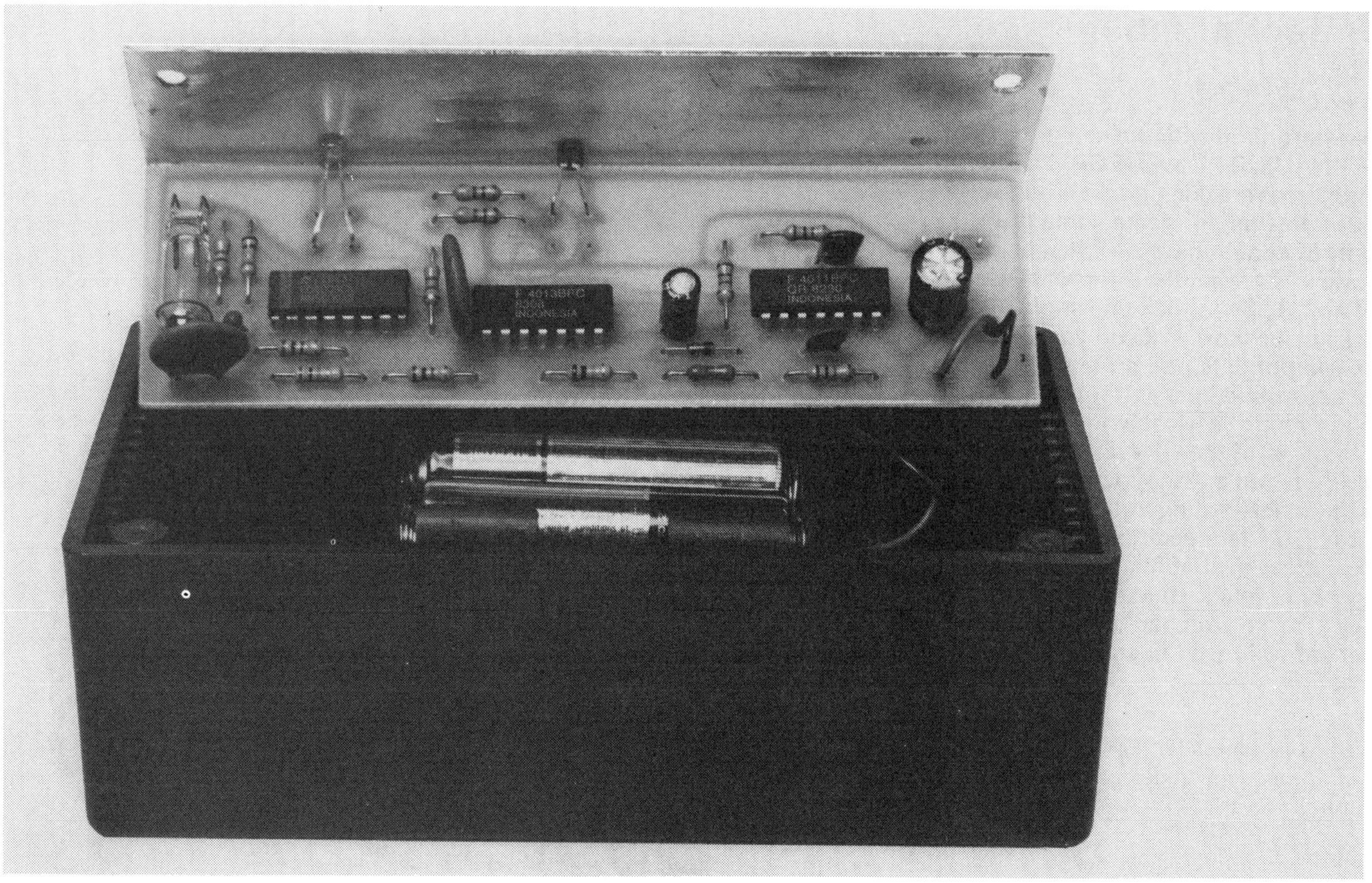

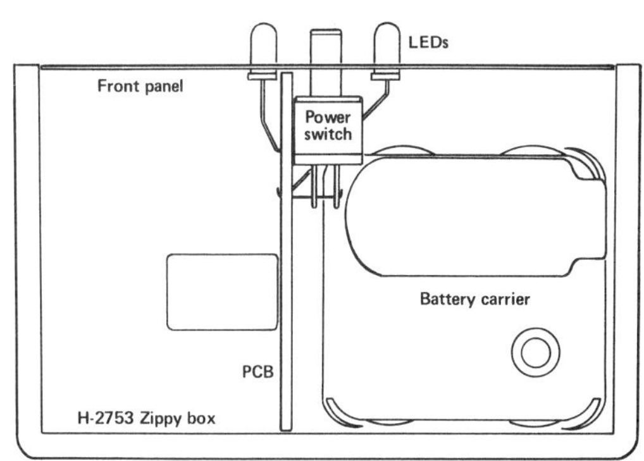

The photograph above shows how the PC board is located with respect to the front panel of the Zippy Box.

The illustration at the right indicates the relative position of the assembly in the box. Notice the PC board is offset from the centre. The battery carrier snugly fits between the PCB and the case.

8. Your near complete PCB can now be slid into the slots provided in the box so that the switch is centred. This means that the board itself is off centre by one slot to accommodate this switch position. Try the lid in place to see that the switch aligns correctly.

9. You can now see how the LEDS have to be positioned and the legs bent to fit the holes in the lid. Note carefully the polarity and the physical position before soldering. Don't cut off the leads till you are assured of correct alignment. You will have to bend leads to achieve the desired position. Be careful not to put excess pressure on the body of the LED, as strain may cause damage and render it inoperative. Use the tip of a pair of long nose pliers if necessary.

10. At this stage, the unit can be checked for operation. Plug in the battery pack and switch on. A flick of the wrist will operate the mercury switch and the LEDS should flash for a short period then stop. If some of the LEDS are not functional, it is possible that they have been installed incorrectly or have been damaged. Recheck and correct as necessary.

11. If all is well, the battery pack and PCB can be installed in the case. This assembly is designed to fit a 6 volt battery carrier along with 4 x AA cells. The P-6124 carrier fits perfectly and should be installed so that the plastic side fits flush with the copper side of the PCB against the solder joints and under the LEDS. This fitting method prevents any movement of the pack and insulates the batteries from the board.

Place the lid in position and screw it down.

Ladies and gentlemen, place your bets.

project number three

Cricket

We started off with the name NOC-TURNALIS STRIDULUS GRYLLIDAE (nocturnal creaking cricket) but decided this might create some problems of communication when trying to describe what the little beast was all about. So to make it simple, it's Cricket for short. With only a handful of components and an input stage based around light and sound sensitive circuits, it closely imitates the behaviour of the GRYLLIDAE family (crickets and grasshoppers). Not only is it an interesting project to construct and get operational, but it can be a great deal of fun as a harmless practical joke! Hidden in a dark, quiet room with an unsuspecting person will result in an entertaining reaction.

Right: The completed Cricket PC Board and wiring to the transducer and microphone.

PARTS LIST

Parts Description		Total Quantity
Resistors:		
1k	R5	1
4.7k	R17	1
8.2k	R3	1
10k	R1, R2, R13	3
22k	R7, R9, R19	3
47k	R4, R6, R10, R16	4
100k	R15	1
220k	R18	1
1 Meg	R8, R11, R12, R14	4
1k	Trimpot VR1	1
Capacitors:		
.0047uF	Ceramic C8	1
.01uF	Ceramic C1	1
0.1uF	Ceramic C3, C4	2
1uF	Tag Tantalum C6	1
2.2uF	RB Vertical Electrolytic C2, C7	2
33uF	RB Vertical Electrolytic C5	1

Parts Description		Total Quantity
Semiconductors:		
4069	CMOS IC 3	1
741	Op-Amp IC 1, IC 2	2
1N4148	Silicon Diode D1-D5	5
Miscellaneous:		
Piezo Transducer L-7022		1
Electret Microphone Insert C-1160		1
Light Dependent Resistor LDR Z-4801		1
Battery Snap P-6216		1
PC Board ZA-1463		1

Parts Description	Total Quantity
Optional Components: (Not supplied in basic kit).	
9V Transistor Battery	1
Case or Model	1
Basic Kit of Components. Cat K-2663	

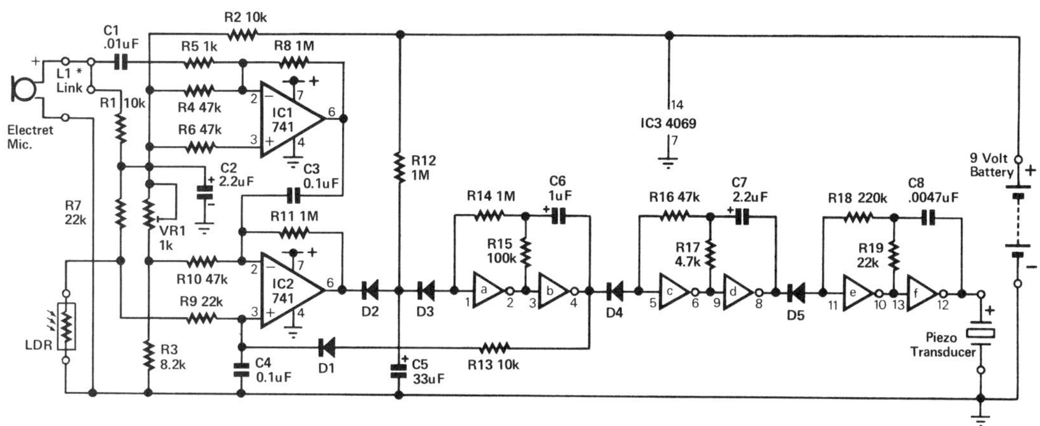

HOW IT WORKS

Under the right input conditions, the circuit produces a chirping noise, not unlike that of a cricket. This output sound will only occur when the ambient noise and light levels are low.

An electret microphone is employed as the sound detector while an LDR (light dependent resistor) is employed as a light measuring device. R1, 10k is a load/supply resistor for this microphone.

IC1 is an operational amplifier configured as a high gain amplifier to increase the low level output from the electret microphone to a figure useable by the second op-amp, IC2. Output from IC1 is AC coupled directly into the inverting input of IC2 via C3. In this case, no input resistor is employed so the AC gain of IC2 is at a very high value while the DC gain is held at around 21 to maintain circuit stability. The LDR is connected to the non-inverting input via a split resistor network.

Let us now look for a moment at the chirping circuits made up from inverters in the 4069 CMOS chip.

In effect, there are three gated oscillators used to make up the chirp sound. The basic frequency is generated by the two inverters, e & f. The capacitor C8 is the main timing component of this astable configuration. Pin 12 is the output point that drives the piezo transducer, the voice box of the cricket. As a

piezo transducer is a relatively high impedance device, a buffer is not required.

Inverters c & d make up another astable that simulates the intermediate frequency that gives the sound its characteristic chirp. When the output at pin 8 goes low, the diode D5 is forward biased and disables the following astable. In this manner, one astable disables the other. C7 is the main timing component of this astable. A change of value gives a difference in frequency and therefore a change to the characteristic of the chirping sound.

The burst time of the chirp is a function of the first gated astable made up of the inverters a & b. Again, the output is coupled via a diode D4 to the input of the following astable. Here the same gating technique is used but in this case, the frequency is very low, around 0.5 seconds. This determines the on-off state of the chirp. This astable is gated via D3 depending on the voltage present on the storage capacitor C5. If this voltage is below the threshold operating point of the inverter a, the astable will be disabled along with the following c & d and e & f. This then is the silent period of the circuit.

Now we can look at the control from the input stages. Firstly assume that there is no noise into the microphone

so the output from IC2 is virtually non-existent.

The input reference for the op-amps is taken to a voltage point under half that of the rail voltage. This is provided by the voltage divider R2 and R3; VR1 allows for some shift in this level.

Now let us assume that the threshold voltage point on pin 1 of the inverter (a) is 6.5 volts (9 volt rail). If for a moment we disregard the influence of the LDR, we can see that any voltage drop across VR1 will be seen by the op-amp IC2 as an offset. The non-inverting input, pin 3 is taken to the more positive end of the pot via R7 and R9. By varying this pot, it is possible to change the output of the op-amp from the reference value to near that of the positive rail. By way of example, we will set VR1 to give a voltage on pin 6 of 7 volt. C5 will now charge up via R12, D2 is reversed biased and D3 is forward biased. As the voltage on pin 5 approaches around 5.8, (threshold of the 4069 of 6.5 minus the 0.7 diode voltage drop), D3 will become reversed biased and the first astable will be enabled and so the chirping will start as discussed previously. Now to the action of the LDR on the circuit. In complete darkness, its value will be in the order of 2 Megohms but in light, the value will drop to around a few hundred ohms. This light value will pull the non-inverting input of the op-

HOW IT WORKS (Cont.)

amp towards the negative rail and the differential voltage will be amplified by a factor of 21. This then causes the output to swing negative and D2 conducts at the point where the voltage crosses the value on C5. This quickly discharges C5 and at the same time, forward biases D3 and so gates off the astables and the chirping. If the LDR returns to darkness, its high value will have little effect on the circuit and so C5 will again charge.

We will now look at the action of the microphone input, assuming the LDR to be in complete darkness.

Noise input to the microphone will be amplified by IC1 and again by IC2. Any negative transition at the output at pin 6 of IC2 will forward bias D2 and therefore discharge C5. The greater the amplitude, the deeper the discharge and therefore the longer the chirp is gated off.

But what happens when the chirp occurs you may well ask? Wouldn't this be 'heard' by the microphone and therefore negate the circuit action? This is where D1 comes into action. When the output of inverter b goes high, as it does when a chirp occurs, D1 is forward biased via R13 and causes

IC2 output pin 6 to go high. This assures that D2 is not conducting and effectively disables the discharging action of C5. C4 slightly extends this pulse to assure circuit stability of the loop.

To set the sensitivity of the total system, VR1 is adjusted in darkness (or a suitable level for normal operation required). The closer the voltage on IC2 pin 6 to the threshold point of IC3a (without sound or light), the greater the sensitivity.

The current drawn by the circuit varies between 1.5mA and 5mA depending on the state of operation.

ASSEMBLY

1. Check off your kit of components against the parts list. Values and quantities should match this list.

2. The components are mounted as shown in the overlay diagram. Firstly mount all the low profile components. Link L1, all resistors and diodes can be loaded into the board. These components can now be soldered after all pigtails are cut to length. Capacitors can next be put into the board and soldered. The last items to be assembled are the pre-set pot and the integrated circuits.

3. The external components are the only components that have to be made to the board. Observe the polarity of these connections. The length of the leads or their position will depend on the housing (if any) into which the cricket is assembled. Imaginative constructors may wish to make a suitable model of a cricket from papier-mache or some form of wire or balsa framework. An on-off switch has not been included but some form of small slider or similar could be connected into the battery leads.

Cricket PC Board
Component Overlay Diagram.

GETTING IT TO CHIRP

Assuming that all the circuitry is correct, there is every chance that your cricket will burst into song, with a little help. Connect a 9 volt battery. If you wish, measure the current drawn by the circuit by connecting a multimeter in series with one of the battery leads. If all is normal, a figure somewhere between 1.5mA and 5mA should be registered.

To test the chirping circuits is a simple matter: simply short out the 1 meg resistor R12. This enables the

chirping circuits as would be the case in normal operation in quiet, dark surroundings.

To set up the front end sound and light detecting circuits, firstly turn the pre-set pot VR1 clockwise (looking at the board as shown in the overlay diagram) After a period, anything up to 30 seconds, the chirping should occur provided that the LDR is in darkness and the ambient noise level is low.

To increase the sensitivity, turn VR1 anti-clockwise slightly. Wait for the result. Several steps will be required to set a satisfactory level. Start this procedure in a completely dark, quiet situation, adjust and then wait. Remember that every time you touch the microphone, the wires or the board connected to it, C5 will be discharged and therefore increase the off period. You will notice that this off period is longer with an increase in the light or noise level.

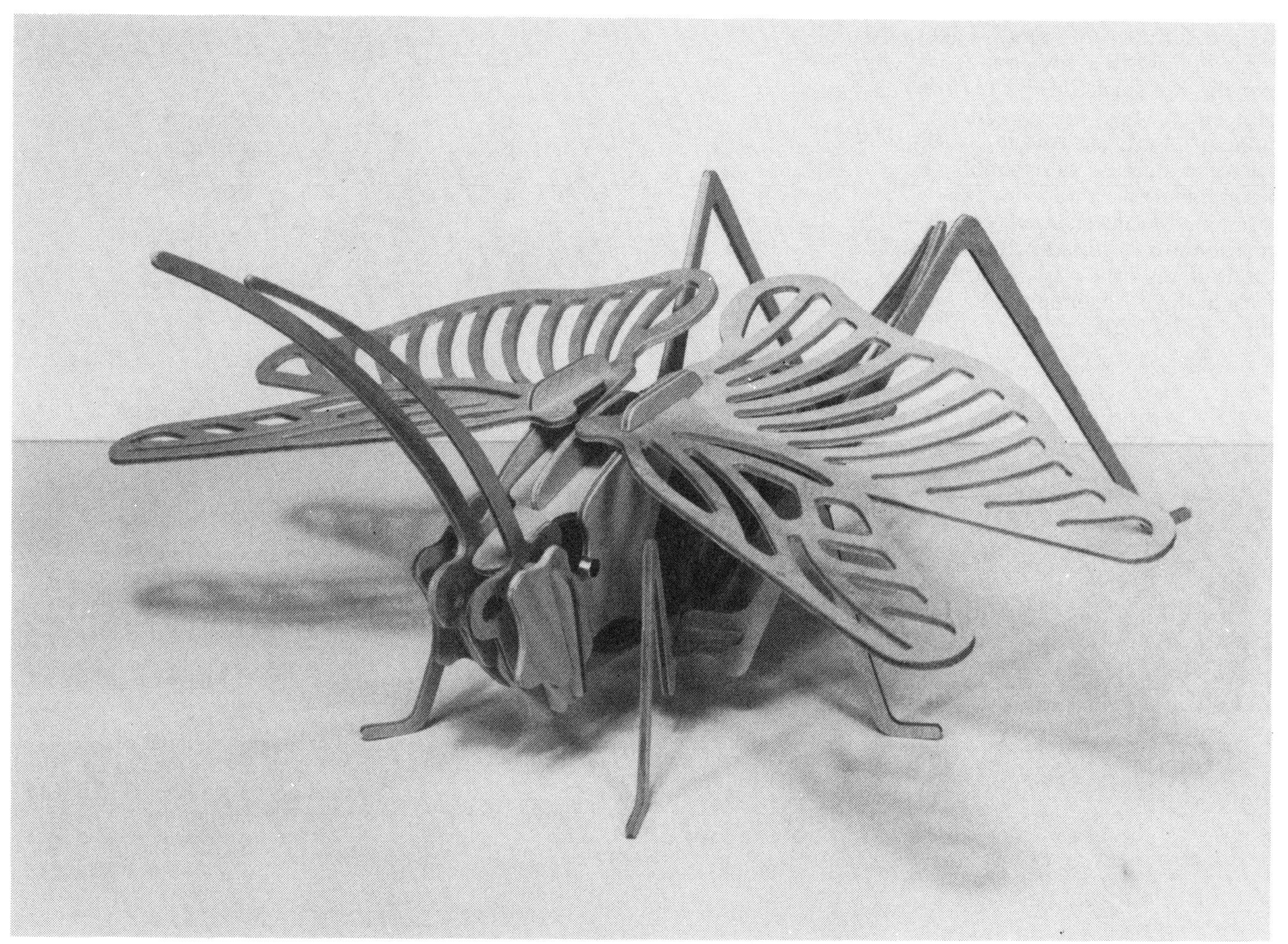

Even without any body (or housing), the completed Cricket Circuit is guaranteed to be a "talking" point at any party. The striking model above is by Tatsuya Kodaka of Japan. The electret microphone insert can be seen at eye level of the insect (Grasshopper in this case).

Variations of Operation

By removing the LDR from circuit and leaving the mic connected, the circuit will not be sensitive to light. In this case it will only be necessary to reset VR1 to obtain the required sensitivity to sound.

By removing the microphone, and leaving the LDR in circuit, the circuit will only be sensitive to light. Note here that this light circuit is very sensitive and even the lowest level will render the cricket quiet. Again it may be necessary to reset the sensitivity.

The cricket can be the centre of a party game or conversation piece by using it in a hide and seek application. In this case, it would require everyone to be completely quiet and the cricket would have to be hidden in a dark place to be effective. Because it is probably not possible to get complete quiet at a party, the sensitivity control may have to be backed off slightly to get the circuit to respond. Even the slightest noise would normally quieten the beast.

Mini Stereo Amplifier

Portable 'Walkie' type stereo cassettes and radios are great for portable entertainment, but sometimes it would be nice if they could be used to give full stereo sound – like a home hi fi. With a few connections to make, this small but high performance stereo amplifier is capable of doing just that, and with top quality sound. To increase its versatility, two preamplifiers have been included to cater for low level input signals – microphones, guitars, turntables, etc. With only a few small modifications to the basic circuit, a PA amplifier can be made.

Above: The completed PC Board Assembly

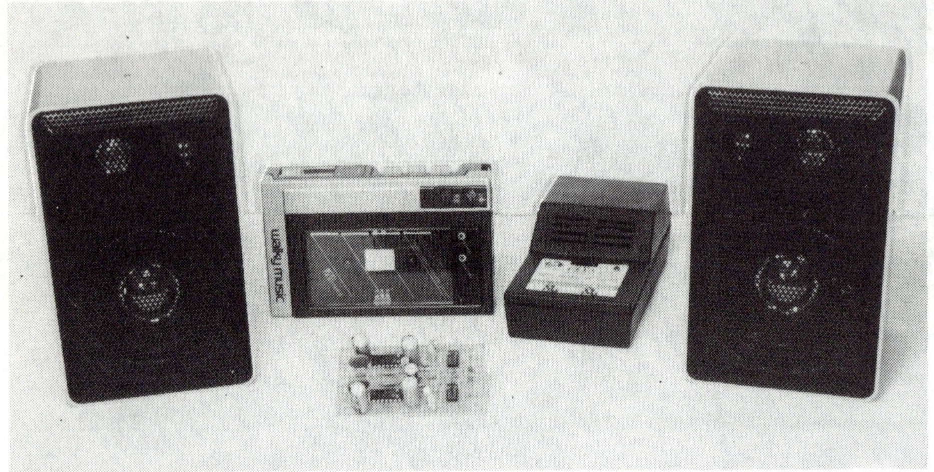

Left: Typical components of a "Big Sound" system showing Speakers, Power Supply, Cassettes, FM Unit and the Mini Stereo Amplifier.

PARTS LIST

Parts Description		Total Quantity
Resistors:		
2.7 ohm	R13, R14	2
2.2k	R1, R2	2
47k	R3, R4, R5, R6, R7, R8, R11, R12,	8
1 meg	R9, R10	2
50k	Log Potentiometers VR1, VR2	2
Capacitors:		
0.1uF	Ceramic C11, C12	2
0.1uF	Tag Tantalum C1, C2	2
2.2uF	RB Vertical Electrolytics C3, C4, C5, C6, C7, C8	6

Parts Description		Total Quantity
100uF	RB Vertical Electrolytic C15	1
470uF	RB Vertical Electrolytics C9, C10, C13, C14	4
Semiconductors:		
741	Op-Amps IC1, IC2	2
LM 380	Audio Amp. IC3, IC4	2
Miscellaneous:		
PC Board	ZA-1467	1
Hookup Wire	500mm	1

Parts Description		Total Quantity
Optional Components: (Not included in basic kit).		
8 ohm speaker systems		2
Power Supply, 12V @ not less than 200mA eg. M-9530		1
33 ohm ½ Watt resistors R-1238		2
47K ¼ Watt resistors R-1116		2
3.5mm Mini Stereo Plug P-1140		1
Hooking Wire – assorted colours W-4010		1
Basic Kit of Components. Cat K-2667		

HOW IT WORKS

The Output Stage:

The LM380 is a small audio amplifier with the internal gain fixed at around 34dB (50 times). A unique input stage allows signals to be referenced to ground. The output is centred somewhere near half the supply voltage. The output is also short circuit proof with internal thermal limiting.

With the supply voltage restraints given and a minimum 8 ohm load, a heatsink on the design shown is generally not required. We have provided a very small amount of heat sinking under the board by using the copper tracks as thermal fins. Although this does not normally represent enough sinking if the chip is to be extended to its maximum capability, in this design and limited parameters, it should satisfy thermal conditions. With a maximum supply of 15 volts and an 8 ohm load, the output is around 1.5 Watts per channel. The input stage is usable with signals from 50mV to 500mV RMS.

The Input Pre-Amplifier Stage:

Two non-committed 741 operational amplifiers have been configured as input amplifiers. Their input stages have been referenced to a common point, that is half the supply voltage. This is simply implemented by using two 2.2k resistors, R1 and R2 to split the supply, C15 is the bypass capacitor at this point. The gain of each of these amplifiers has been fixed at 21 by the input resistors R3, R4 and the feedback resistors, R9, R10. Input capacitors, C1,C2 isolate any DC component from the signal. The output point has been taken to a connection pad on the board to allow for maximum user flexibility. In most cases, this would be used to drive the final stages via capacitors, C3, C4 and volume controls, RV1, RV2. If required, these stages could be used separately, maybe as experimental bench amplifiers. In one application shown later, they are configured slightly different to use the board as a PA amplifier.

With a power supply of 12 volts, the quiesent current drawn by the total system is 30 to 35mA, under driven conditions, the drain could increase to 300ma or more.

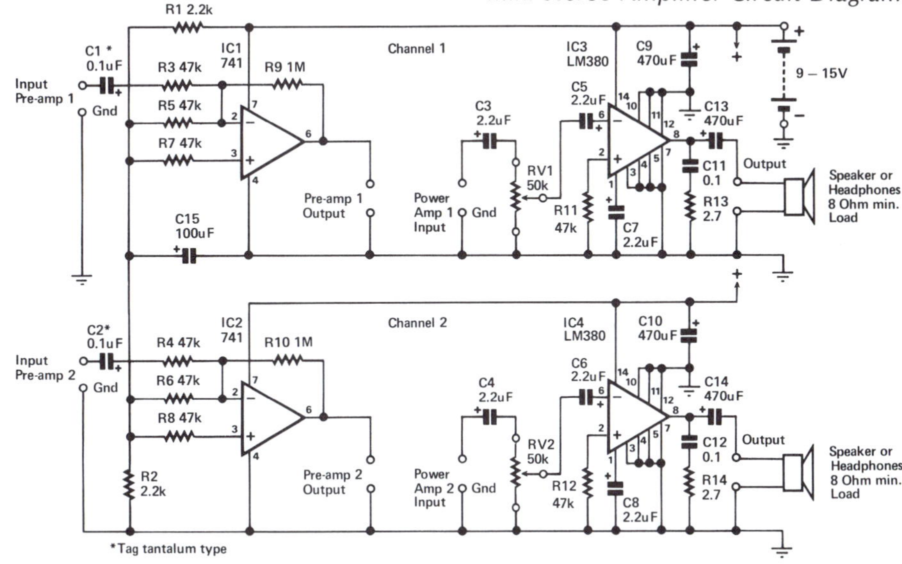

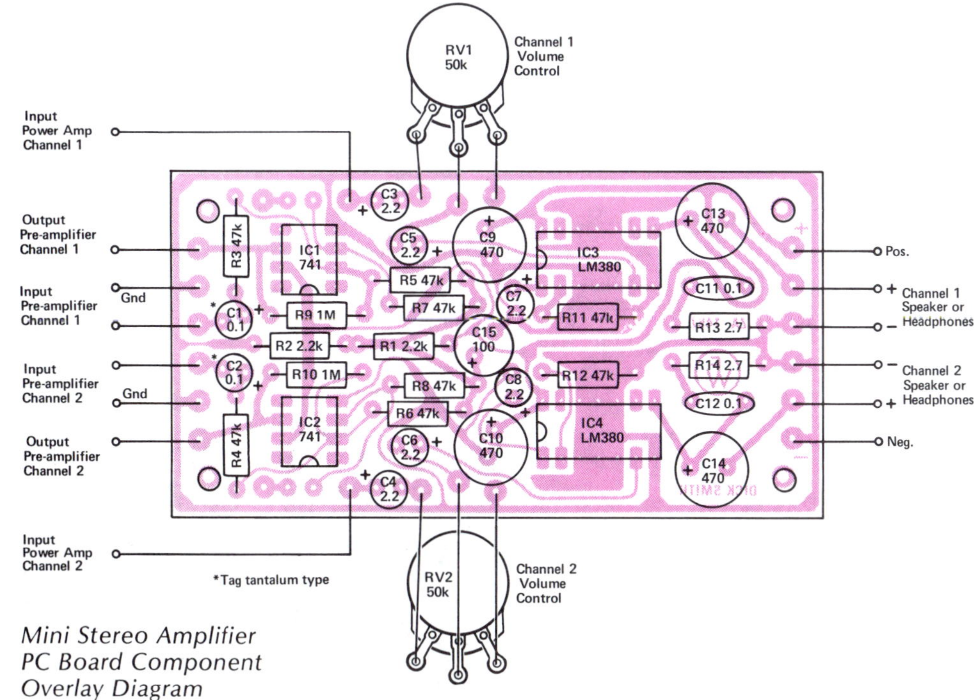

Mini Stereo Amplifier
PC Board Component
Overlay Diagram

ASSEMBLY

1. Check off the components in the kit against the parts list.

2. Insert all the resistors, cut pigtails and solder.

3. Next insert the ICs and solder. This may appear a little unconventional but because of the size of the capacitors that flank these ICs, it is easier to load in this sequence.

4. Now the capacitors can be inserted, trimmed and soldered. Note carefully the polarity of electrolytics.

5. The only connections left are the volume controls, RV1 and RV2, the input wiring and the power wires. The input wiring will be dictated by how you use the system. In a case where you are to configure the amplifiers to provide "big sound" for your portable Sound Tripper type AM/FM Stereo System, only the output stages are used. Where a low

level signal, like a microphone input for example is to be used, the pre-amplifiers would have to be included.

6. When all connections are checked and a final inspection of the soldered joints under the PCBs have been made, power can be applied. For a power supply, you will need at least 9 volt at 200mA. The Dick Smith M-9525, M-9514, M-9560, M-9526, M-9550 and M-9530 would all be suitable. Obviously, the 12 volt supplies would give the amplifiers the greatest audio output capability.

BIG SOUND FROM YOUR PERSONAL WALKING TYPE STEREO

Normally this type of miniature AM/FM or cassette radio player is only used for personal listening with headphones. Now you can get the big freedom sound by using your Mini Amp as the stereo output stage, simply plug it into the headphone socket.

By following the diagram, you will see that any type of 8 ohm speakers will suit. The power supply is not critical but should be within the 9 to 15 volt limits. A nominal value of 12 volts would be ideal.

The input to the amplifier is connected to the output section only, the front end 741 pre-amps are not used in this case.

The headphone connector to these portable type stereo units is normally a miniature type 3.5mm plug (the Dick Smith P-1140 for example).

Simply solder one of these plugs to the three wires as shown, one for each channel, with a common earth. The common is the body or the long stem of the plug.

The 33 ohm resistor connected across each channel to ground is a dummy load for the radio player when headphones are not connected. Where the unit has two headphone output sockets, as most do, and you can leave one set of phones connected, these two resistors are not necessary. In some types, the headphone plug is the on-off switch.

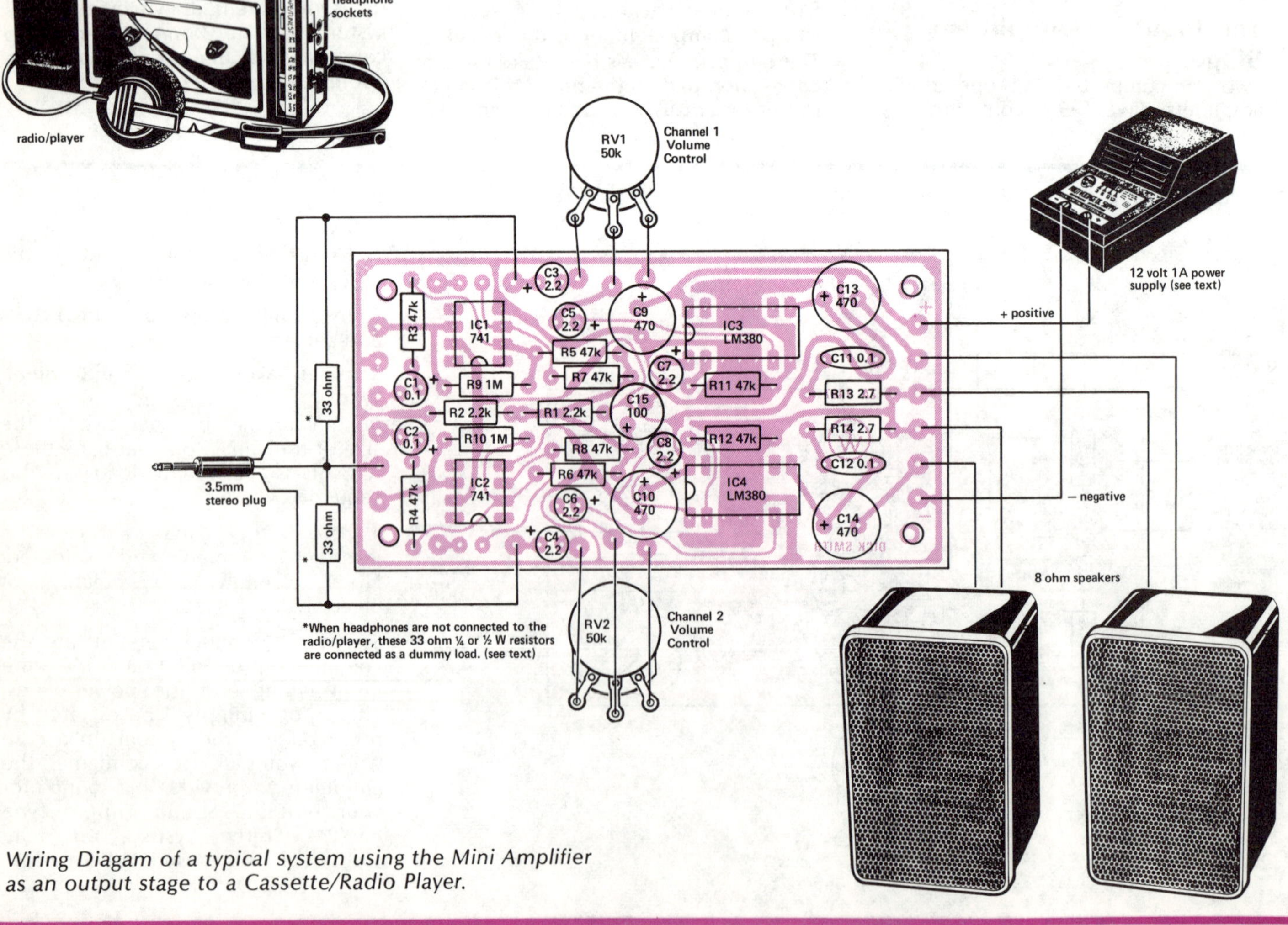

Wiring Diagam of a typical system using the Mini Amplifier as an output stage to a Cassette/Radio Player.

USING PRE-AMPLIFIER STAGES FOR EXTRA GAIN

The output of the pre-amp is connected to the input of the output stage as shown. In this configuration, the system could be used as an experimental amplifier where low level signals are present. It makes a handy bench or workshop amplifier. Although no special frequency response tailoring has been included around the 741 stages, it could be used as a microphone amplifier using dynamic or electret types. These stages are usable with input signals from 3.5mV to 100mV RMS.

SMALL PA AMPLIFIER OR MEGAPHONE

By using the twin output stages in a "bridge" mode, the output power can be approximately doubled.

As the diagram shows, the speaker is connected across the active output points of each amplifier. The earth or common in this case is not connected.

Pre-amplifier 1 is used as an input stage with a gain of 21 and is useful with signals from 3.5mV to 100mV. The second pre-amp has a different function in this case. The gain has been reduced to unity by changing the feedback resistor R10 to 47K. Now this stage becomes an inverter of the signal to the second output stage. This satisfies the requirements of the bridge output in that one input is positive going while the other is negative. In other words, the input to the output stages are 180 degrees out of phase. This provides twice the voltage swing across the 8 ohms load for a given supply thereby increasing the output power by a factor of four over that of a single stage. However, the package power dissipation will be the first parameter limiting power delivered into the load. You see that in this case we have limited the power supply to a maximum of 12 volts. In all, the final result is an output power capability of around double that of a single amplifier. The configuration shown is suitable for input signal levels from 3.5mV to 100mV RMS. Both the volume controls RV1 and RV2 have to be turned up and down by the same amount to control the output.

If you wish, for the sake of convenience, RV1 and RV2 can be replaced by a dual gang pot.

Where high gain is not required, the circuit has to be changed. As with the modified stage of pre-amp 2, the feedback resistor R9 of pre-amp 1 has to be changed to a value of 47k so that it only has unity gain. In this way, higher input signals from 50mV to 500mV can be handled. This makes the system ideal as a "big sound" amplifier for your mono radio or cassette player. The connections can be made to the speaker or earphone sockets as described for the Big Sound Stereo Amp.

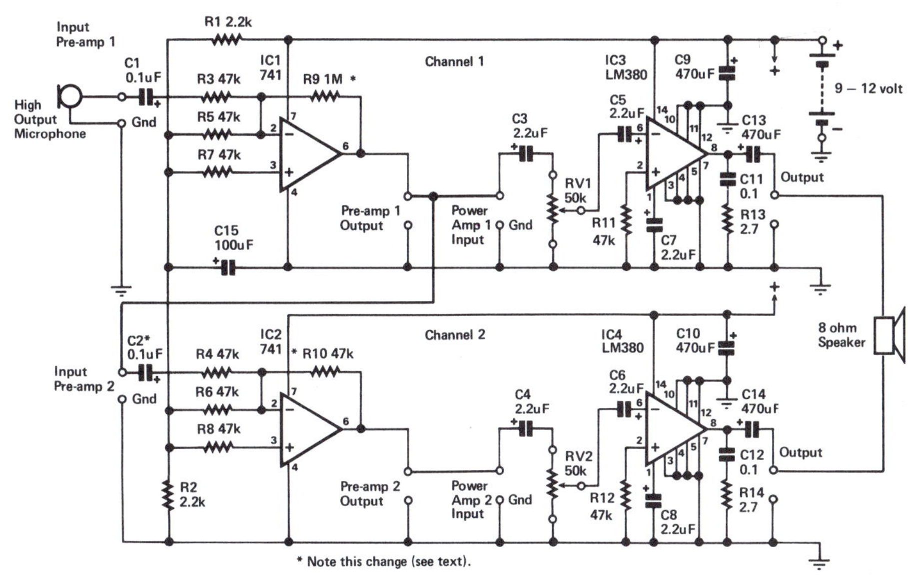

Circuit Diagram of PA Amplifier

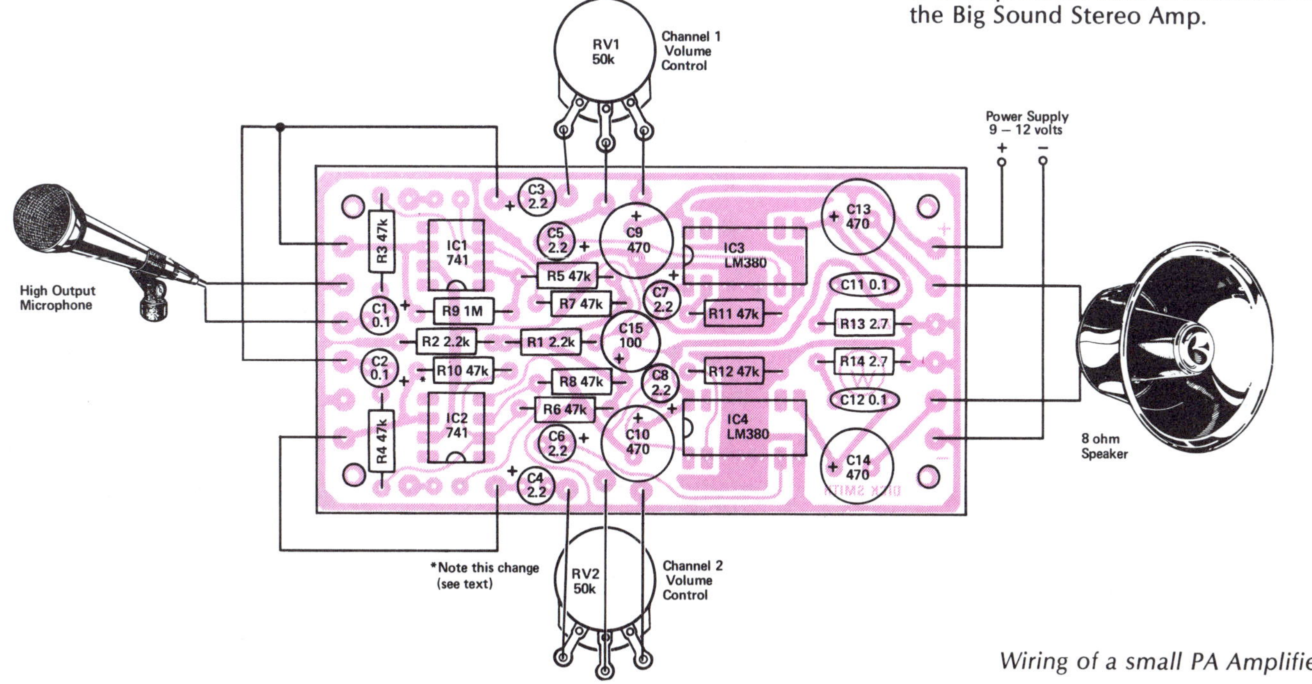

Wiring of a small PA Amplifier.

Minder

This automotive based project is aimed at the ever increasing problem of vehicle theft. It may not always be obvious that an alarm system is fitted to a car and an attempt to illegally enter it by unwanted villains leaves many people feeling uneasy about their security. In many cases, the additional inconvenience of broken windows, locks, etc only adds insult to injury. Not only does this little device give a flashing warning to would-be intruders but also aids the driver or occupants under normal use. Doors left open, or opened while driving gives an alert signal. It's great with kids in the back seat! If you walk away from your car and leave the lights on, this circuit will give a warning. No more flat batteries.

Left: The final assembly in the optional Zippy Box. Although all components are shown mounted in this photograph, you may wish to have the switch, LED or lamp at a remote point.

Right: The completed PC Board of the Minder circuit prior to wiring.

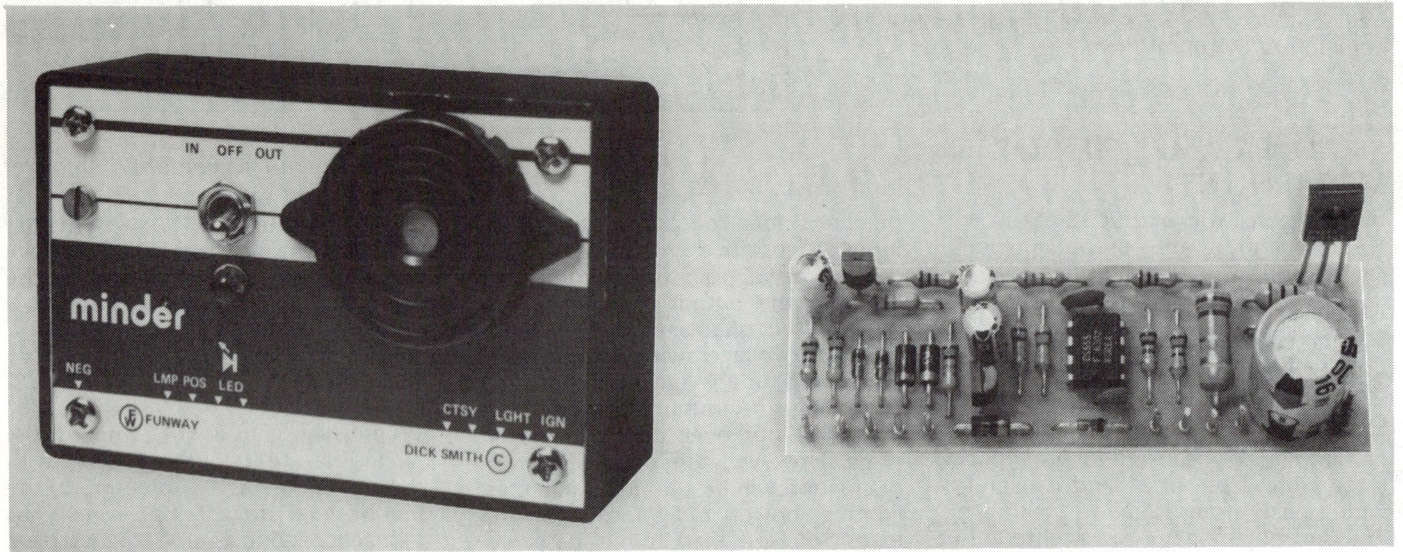

PARTS LIST

Parts Description		Total Quantity
Resistors (all ¼W unless stated)		
2.2 ohms	1 watt R14	1
10 ohm	½ watt R11, R12	2
270 ohm	R13	1
1k	R3, R4, R9	3
1.5k	R10	1
10k	R1, R2, R6	3
33k	R5	1
100k	R8	1
220k	R7	1
Capacitors;		
.0022uF	Ceramic C4	1
.01uF Ceramic	C5	1
2.2uF RB Vertical Electrolytic	C1, C2, C3	3
1000uF RB Vertical Electrolytic	C6	1

Parts Description	Total Quantity
Semiconductors:	
555 Timer IC IC1	1
BC548/DS548 or 2N2222 NPN Transistor TR1,TR2	2
BD140 or 2N4920 PNP Transistor TR3	1
1N4148 Silicon Diode D1, D2, D6	3
1N4002 Power Diode D3, D4, D5	3
5mm Orange LED LD1	1
Miscellaneous:	
Piezo Transducer L-7022	1
PC Board Pins	12
PC Board ZA-1460	1

Parts Description	Total Quantity
Optional Components: (Not supplied in basic kit).	
12 Volt Red Bezel S-3510	1
Centre Off Switch DPDT S-1286	1
NOTE: The two above components can be replaced by one device – Illuminated Rocker Switch S-1503	
Hookup Wire – 5 colours 10 x 0.2mm or 24 x 0.2mm, lengths to suit vehicle wiring. Pack.	1
Terminal Strip H-6700	1
Zippy Box Case UB5 H-2755	1
3mm/6BA x 12mm Screw 3mm/6BA Nut. H-1684 pack	1
TO-220 Mica Washer H-1916	1

Basic Kit of Components. Cat K-2660

HOW IT WORKS

The circuit consists of three main sections: the high/low frequency oscillator centred around the 555 integrated circuit; the output stage with TR3, LED LD1 and the piezo transducer; the input gating stage consisting of TR1, TR2 and the diodes D1 to D6.

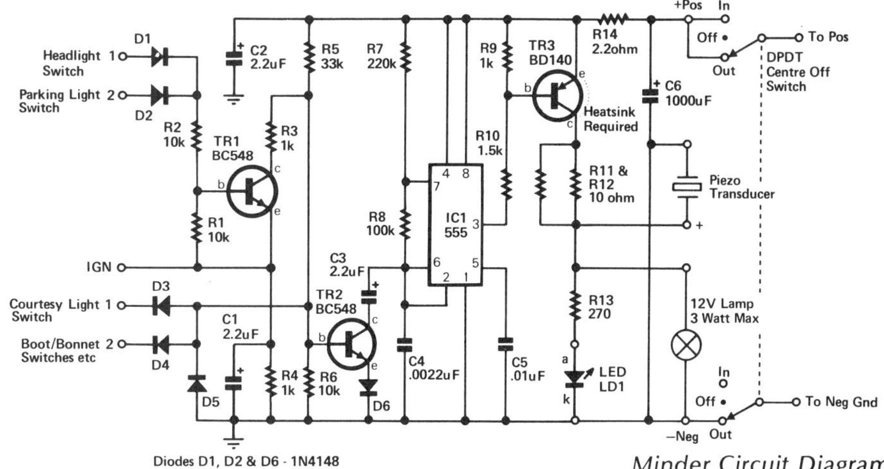

Minder Circuit Diagram

Astable

The heart of the system is a 555 timer integrated circuit configured to operate in the astable mode. When in the normal state, TR2 is switched on and therefore C3 is effectively in parallel with C4. This sets the frequency of oscillation to around 1Hz, the lamp flash rate. If TR2 is off because of the state of the gating circuits, C3 will be out of circuit and therefore only C4 will determine the frequency. In this case, the frequency is around 1350Hz, the alarm tone heard in the transducer. This transducer also emits a low click at the 1Hz rate.

Output

The output from the 555 is directly coupled to the output transistor TR3 via R10. When this output at pin 3 goes low, (for about one third of the duty cycle) the PNP transistor is switched on and therefore sources current for the load. R13 limits this current to around 35mA for the LED. The two parallel resistors R11 and R12 are used to limit the in-rush current of the lamp. If a low current indicator like the S-3510 red bezel is used, the on current in the output stage is approximately 100mA and well within the limits of the BD140 transistor. If a higher wattage lamp is used, (3 Watt max) the on current may be within the parameters of the transistor but the in-rush current to the lamp may exceed these values. To be on the safe side, we have indicated that this power transistor should have a heatsink. The lid of the Zippy Box in which the PC board can be housed is ideal. If this mounting arrangement is not used, a small piece of aluminium sheet of around 1250 square millimetres can be used.

Components R14 and C6 are included to suppress transient currents from the 555 and the in-rush current to the lamp. Where a current sensing burglar alarm is fitted to a vehicle, the power needed by the flashing system may be enough to trigger this alarm if these two components were not fitted. In a case where the burglar alarm is very sensitive and the flash lamp is the maximum of 3 Watts, there still may be a tendency to trigger the system. In this case, it may be necessary to increase the size of the capacitor C6 and the value of R14. Try 2500uF for C6 and 4.7 ohm for R14.

Gating

The gating circuits operate in three states on the change of polarity at the input.

Firstly, the courtesy light sensors with switches that operate against negative ground. Normally with all switches off, the input to the diodes would be positive or open circuit if the courtesy lamp is not in circuit or blown. In this case, the diodes would be reversed biased and have no effect on the TR2 circuit. TR2 would be forward biased via R5, assuming that TR1 was not switched on. C3 would be switched into circuit to produce a 1 second flash rate. Now assume that one of the door switches on the courtesy light system is switched on. The diode D3 or D4 depending on which one is connected, is forward biased via R5 and this deprives TR2 of base current and switches it off. The diode D6 is included in the circuit to lift the base/emitter voltage above negative ground to assure correct operation of this gating action. C3 is now out of circuit and as a consequence, the astable runs at the high alarm tone rate.

Secondly, the diodes D1 and D2 together with TR1 operate against positive in the active state. Assume that the ignition switch is off. This connects the emitter of TR1 to negative ground through the components of the electrical system of the vehicle that are normally powered via the ignition switch when it is on. The ignition coil, (if the points are closed) the battery charging system and the instrument circuits are part of this path to ground.

If the headlights or parking lights are on, the diodes D1 and D2 are forward biased via R2 and the base/emitter junction of TR1. This turns TR1 on and again deprives TR2 of base current. The same action as described previously then follows.

When the lights are off, the diodes and TR1 are non-conducting and this circuit has no effect on TR2.

If the ignition switch is on, TR1 is turned off and therefore does not gate TR2. Even though the lights may be on in this state, no alarm is required as this condition would be normal with night driving.

C1 is included to suppress transients on the ignition power line. R3 provides current limiting in TR1 to Zener breakdown voltage that occurs when the emitter/base is positive with respect to the collector.

Thirdly, when the negative power wiring to the unit is switched off, positive still active, the circuit will be disabled. This is shown on the wiring diagram with the power switch in the 'in car' position. This state is used to warn the occupants of the vehicle that a door, bonnet or boot has opened when driving. As soon as one of the switches connected to these points makes contact to ground, the diodes D3 or D4 together with D5 connects the negative line of the circuit. This circuit action also makes sure that TR2 is off and therefore causes the astable to run at the high alarm rate.

If the ignition is off in the 'in car' position, a low level alarm tone will be heard. Resistor R4 provides a low current path to ground to cause this action. This is a reminder to the driver that the unit is still on and has to be set to the 'out of car' position or switched off.

ASSEMBLY

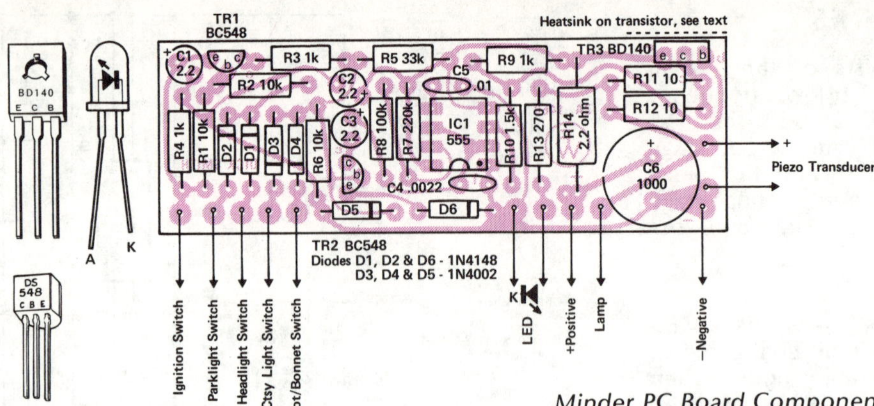

Minder PC Board Component Overlay Diagram

1. Check off your kit of components against the parts list.

2. Load all low profile components including resistors and diodes into the board. Note the polarity of the diodes. Cut pigtails and solder in position.

3. Insert all capacitors, again being careful of polarity. Cut and solder. Insert PC board pins in the output wiring point holes and solder.

4. Insert the 555 IC and the two BC548 transistors. Do not insert and solder the BD140 and the LED at this stage.

5. The housing or mounting of the PC board will now have to be considered as this will affect the wiring. If you are not going to use the suggested Zippy Box mounting method, carefully plan the wiring and mounting arrangement.

6. The PCB is designed to fit into a UB5 H-2755 Zippy Box. The front panel label for this box can be cut out from the rear of this book. The position of the external components to the board are shown on this panel but it is not necessary that they are mounted in this way. You may wish to use only the LED as the flashing indicator and mount it in some prominent position in the dash or instrument panel of your car. The power switch can also be mounted in the dash. These differences will have to be thought about in your wiring plan.

7. To position the components on the Zippy box lid, simply place the label accurately on the surface and mark the centres with a sharp scriber or simliar instrument. Drill all holes as required--the S-1286 centre off switch requires a 6mm to 6.5mm hole, the LED 5mm, the transistor 3mm and the transducer wiring access hole can also be 3mm. If you wish to mount the transducer with screws, drill 2 x 2mm holes at the position shown. The label can now be glued onto the lid with a suitable rubber based contact adhesive. Punch out the holes by using the shank of the previously used drill. Push it through from the front to rear surface. A hole will also have to be drilled in the centre of the bottom of the box to allow wire access. The size and the number of wires coming from the board will dictate the size of this hole; around 6.5mm will generally be large enough.

8. If you wish, a terminal strip such as the Dick Smith H-6700 could be used as an easy means of connection. This could be mounted on the rear of the Zippy box. This means that the wires to the PCB would only have to be short. Alternately, if the wires are to go directly to the external connections, they have to be long enough to go from the box mounting position.

9. Before the wiring is soldered, the BD140 transistor has to be aligned, soldered and mounted. This is done by sliding the PCB into the last slot position in the box. Now loosely fit the transistor into its holes in the PCB the right way up with the metal section facing towards you. Now place the lid above the normal fitted position on the box and move the transistor around till the mounting holes align. Without moving this position, solder the transistor legs. If the LED is to be mounted in the panel, it can be positioned and soldered to the PC board pins. The legs may have to be bent slightly to accommodate the body in the hole.

10. **PCB Wiring.**
All wires can now be soldered to the appropriate points on the PC board. Be careful to see that all strands of the wires are around the pins. Stray strands may short out against adjacent components. If the power switch to be used is the S-1286 and is mounted in the box as shown, it can be connected with short lengths of wire to the PCB as necessary. The piezo transducer can be mounted on the lid and the wires brought through the hole provided. It can either be stuck to the surface with glue or double sided tape or bolted using 2mm/8BA screws. The wires are soldered to the appropriate points.

11. The unit can be tested at this stage. Connect the 12 volt power supply. Without any of the gating wires connected, the LED should flash when the switch is in the 'out' position. A low level click should also be heard from the transducer. Now touch one of the courtesy wires to the negative terminal; the alarm tone should sound. The LED or lamp, if connected, will glow at low level. Now connect the IGN wire to negative and then touch one of the LGHT wires to the positive line. Again, the alarm should sound. Flick the power switch to the IN position. If the IGN wire is connected to negative, a low level tone will be heard in the transducer. Now touch a CTSY wire to negative; the full alarm sound should be heard. If all is well, the case assembly can be finished. If all or some of the functions do not operate as indicated, the wiring, connections or circuit will have to be checked.

12. Mount the power switch to the panel. The power transistor can be attached with a 3mm/6BA screw. Do not forget the mica washer between the body and the lid. If you wish, thermal compound can be smeared on either side of this washer to improve heat transfer efficiency although it is really not necessary in this application. Do not tighten the screw securely at this stage. The LED should fit snugly through the mounting hole. The total assembly can now be slid into the last slot in the Zippy box. Be careful not to pinch out any of the wires under the PCB. When the assembly and the lid are mounted squarely in position, tighten the transistor mounting screw and fit the case screws.

WIRING THE CAR

1. The positive lead is connected via the power switch to a permanently active positive point in the vehicle's wiring system. Usually this can be found on the fuse box.

2. The negative lead is connected to the frame of the vehicle via the power switch as shown.

3. If an external lamp is used, this can now be wired. If you have decided to use the LED as the warning lamp, the wiring and mounting can be carried out. It would be wise to put some spaghetti insulation over each leg to avoid short circuits. The mounting of the

Components mounted
inside optional Zippy Box.

LED is simple: all that is required is to drill a 5mm hole through the panel in which it is to be mounted. It can be held in place with a small amount of glue. If you wish, a small mounting bezel like the Dick Smith H-1910 can be used.

4. The wire to the headlight must be connected to the lamp side. This point may not always be easy to find and some experimenting may be required. In a case where the parking lights do not come on when the headlights are on, both input wires will have to be connected to the lamp sides of the switch. This makes sure that both states are covered, that is, when the headlights are left on or the parking lights are left on.

5. With the courtesy dome light system, the wiring is taken to the door switches. Again, this wire is connected to the lamp side. If you wish, the second diode of this sensing circuit can be taken to a boot or bonnet switch as a warning that this compartment is not shut properly.

6. If you have decided to use an external power switch, this can be mounted and wired. The box itself will have to be mounted in some convenient location by screws through the side.

After wiring is complete, the total assembly can be tested.

ALARM WARNING LIGHT

This project can be used as an alarm warning light only. Irrespective of whether an alarm system is fitted to a vehicle or not, the fact that this flashing system is fitted and is visible from outside the car should deter any would-be intruder.

The external wiring to the circuit is much simpler in this case. Only an on-off switch in the positive line is required. The negative wiring is connected directly to the chassis (negative) of the car. None of the front end gating circuits are used.

If you wish, all gating components up to and including TR2 can be left out of the circuit. The only difference is that C4 will have to be replaced by a 2.2uF capacitor (C4) to affect the 1Hz flash rate. Either or both the LED and Lamp can be used as the indicator. The transducer is not used in this case.

Minder Wiring Diagram showing
connections to vehicle components.

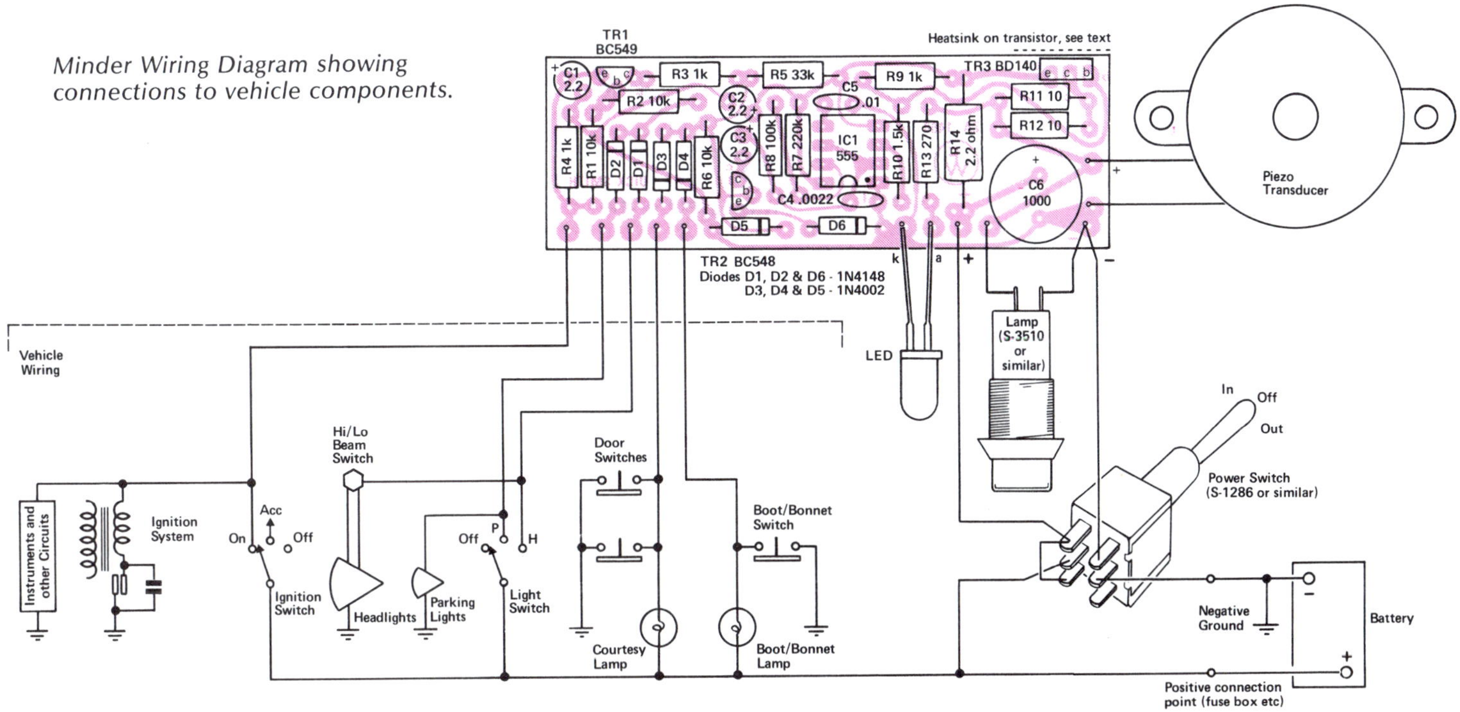

Mini Colour Organ

This simple but effective circuit gives an extra dimension to your musical entertainment. Simply connect to your radio, cassette or stereo for a bright, solid state LED display. With only a single IC and a handful of discrete components, this project can be constructed in a few hours. Operating over a wide voltage range, it suits a large number of signal sources. Two units can be constructed for stereo signals, or the stereo signals can be added together to give one display.

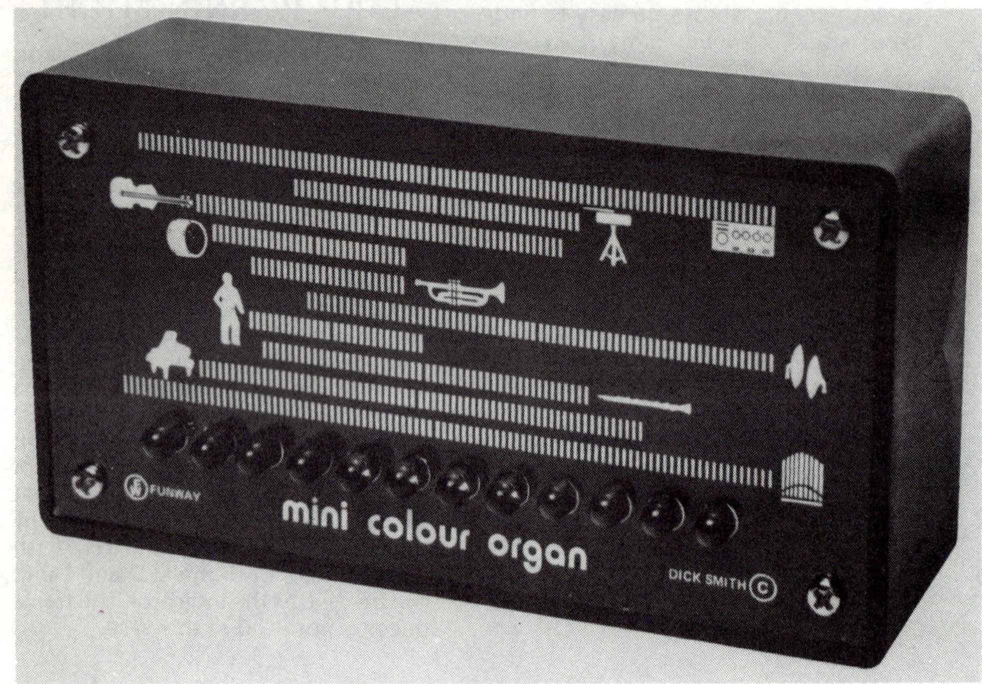

The completed unit shown in optional Zippy Box.

PARTS LIST

Parts Description		Total Quantity
Resistors:		
10 ohm	R20	1
100 ohm	R21	1
470 ohm	R3, R4	2
10k	R9, R10, R11, R12, R14, R18, R19	7
33k	R6, R7	2
39k	R2, R8, R15, R16, R17	5
47k	R1, R13	2
1 Meg.	R5	1
200 ohm	Trimpot RV2	1
500 ohm	Trimpot RV3, RV4, RV5	3
50k	Trimpot RV1	1
Capacitors:		
.0033uF	Greencap Polyester C6, C7	2
.0068uF	Greencap Polyester C8, C9	2
.056uF	Greencap Polyester C12, C13	2

Parts Description		Total Quantity
2.2uF	RB Vertical Electrolytic C1, C5, C10, C11	4
100uF	RB Vertical Electrolytic C2, C3, C4	3
470uF	RB Vertical Electrolytic C14	1
Semiconductors:		
	LM 324 Quad Op-Amp. IC1	1
	1N4148 Silicon Diode D1, D2, D3, D4	4
	7.5V, 1W Zener Diode ZD1	1
	5mm Green LED's LD1 – LD4	4
	5mm Orange LED's LD5 – LD8	4
	5mm Red LED's LD9 – LD12	4

Parts Description		Total Quantity
Miscellaneous:		
PC Board	ZA-1464	1
Hookup Wire – 1m.		1
Optional Components:		
(Not supplied in basic kit).		
Zippy Box Case UB3	H-2753	1
Hookup Wire – assorted colours	W-4010	1
50k Log or Lin pot.	R-6823 or R-6809	1
Power Supply, 12 volt @ 200mA min.	eg. M-9530 or M-9526	1
Basic Kit of Components.		
Cat K-2664		

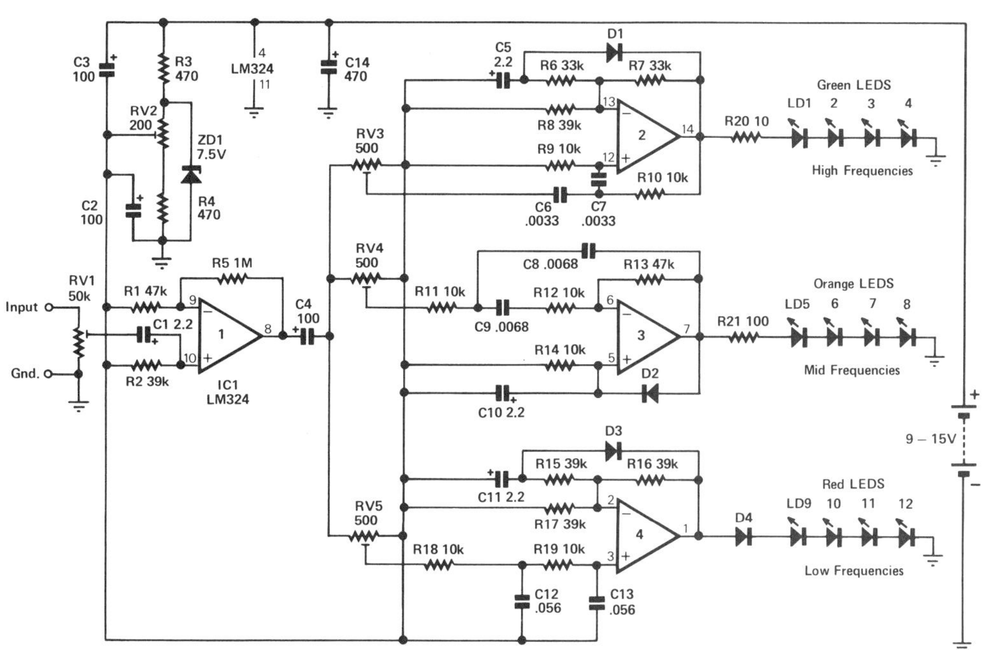

Mini Colour Organ.
Circuit Diagram.

HOW IT WORKS

The total circuit is based on a quad operational amplifier, LM324. One section is used as an input amplifier to raise the signal level to a figure usable by the following three active filter circuits. The output of these filters directly drives the display section, three banks of four LEDS.

The input stage of all op-amps is taken to a common adjustable reference point. Zener diode ZD1 stabilises this voltage for input supply rail voltages ranging between 9 and 15 volts. RV2 adjusts this reference point around 6.5 volts to be used as the threshold point to switch on the LEDS.

The input signal is fed into the pre-amplifier (1) via a 50k pre-set potentiometer that acts as a master level control. The signal is amplified around 20 times by this stage and fed to each filter via individual level controls RV3 to RV5. Op-amp 2 is the high pass filter, 3 is the mid band pass filter and 4 is the low pass. The filters themselves are reasonably basic.

The low pass, section 4 is a modified first order design. Op-amp 3 is used in a multiple feedback band pass configuration. A modified second order filter is used as the high pass. As the actual performance of each filter does not have to meet any strict criteria, no attempt was made to choose precise values in order to achieve text book figures. The low pass operates up to around 500 Hz, the band pass operates in the region from 1kHz to 2.5kHz while the high pass is active in the 2.5kHz to 15kHz region. These figures would tend to indicate that there are 'holes' in the response but in actual use, the controls are easy to setup and this results in a smooth changeover and dynamic display.

The diodes D1 to D3 in each filter are used to make the display appear brighter. Where the signal or music is of staccato nature, the display would tend to be dull. This is not because the LEDS are not driven hard enough but because the on time is very short. This short burst of light is interpreted by your eyes and brain only as a dull blink. This is where the diodes and storage capacitors C5, C10 and C11 come into action. The combinations act as pulse extenders so that the on time of the LED is increased. In this way the brain is effectively fooled into registering the sensation as a short , bright burst of light.

Resistors R20, R21 and the diode D4 are included in the display to equalize the threshold switching points and brightness of the LEDS. The operating voltage of LEDS varies slightly with different colours, red around 1.6V, orange around 1.7V and green around 2.2V. This is the reason for the odd differences at the output stage of each op-amp.

The standby current drawn by the circuit is approximately 13mA with a DC rail of 12 volts. The power supply used with this project should be capable of handling 200mA or greater.

Input Voltage: 100mV PP minimum value for satisfactory operation, 400mV PP for smooth operation.
Input impedance: 20k
DC Supply: 9 volt min. to 15 volt max. @ 200mA min.

ASSEMBLY

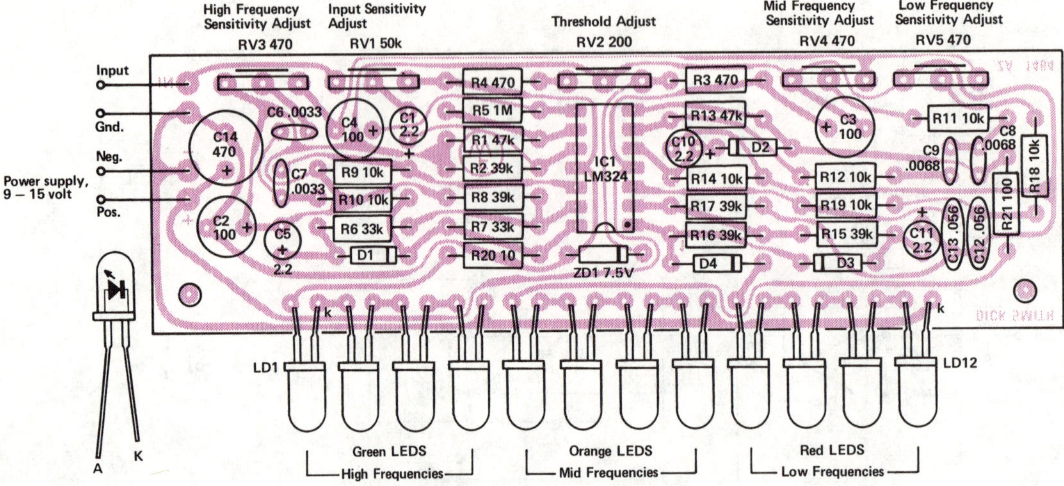

PC Board Component Overlay Diagram.

1. Firstly check off all the components supplied against the parts list.

2. Next load in all the low profile components including the resistors and diodes. Cut off all pigtails and solder the joints. Now all capacitors can be installed and soldered. The pre-set pots followed by the LM324 IC are then inserted and soldered.

3. The last components to go in are the 12 LEDS. How you wish to house or mount the board will dictate how the legs of the LEDS are to be bent. The board will slip in lengthways into a H-2753, UB3 Zippy Box. In this case the LEDS would have to be bent at right angles to penetrate through the lid. Be careful when making these bends: don't put excessive strain on the body. Hold the legs near the body itself with the tips of a pair of long nose pliers.

To mark the LED centres in the Zippy Box lid, firstly cut the label from the page at the rear of this book. Place it accurately over the lid and mark the 12 centres with a scriber or similar sharp pointed instrument. Now drill 5mm holes at these points and clean off all burrs. With rubber based contact adhesive, glue the label accurately onto the lid. The rear shank of the drill can be used to punch out the holes in the label by pushing through from the front surface.

4. All that is left to do is connect the necessary wiring to the power supply points and to the input stage.

5. Where the Zippy box has been used to house this project, some provision has to be made for the wiring. A hole can be drilled through the rear bottom of the case to allow for access. 4mm to 6mm will suffice.

To allow the five controls to be adjusted with the board inside the case, holes will have to be marked and drilled at the adjustment point centres. Once set, four of these controls don't really require further adjustment. The only control that does have to be continuously set is the Input Sensitivity adjust pot RV1. If you wish, the preset control can be replaced by a conventional pot with shaft and knob. It could be mounted at the rear top of the case with access from the back. Three wires would be necessary to connect it to the PCB.

Getting It Operational

The nominal power supply for this unit is 12 volts although any value between 9 and 15 volts can be used. When the power is first applied, the LEDS will give a momentary flash. This is a rough indication that at least most of the circuit is operational. The next quick test is to measure the current drawn by the circuit by connecting your multimeter in series with the power supply. With no input signal, a reading in the order of 10 to 25mA should result with a 12 volt DC supply. A reading well outside this area would tend to indicate that something is wrong.

Now assuming that all is correct, adjust the threshold pot, (RV2) clockwise (when looking at the board as shown in the overlay diagram). If you still have your meter connected, you will see that the current increases as the pot is turned. If all is normal, some of the LEDS will start to light. The correct setting for the pot is just before the point where the LEDS give a faint glimmer.

Now connect some form of audio signal to the input. It may be from the speaker or earphone terminals of a transistor radio or the output of an amplifier. The only requirement is that the signal is over 100mV PP in level. The radio or amplifier turned to the normal listening level will do fine. Next turn RV1, the input sensitivity control anti-clockwise till there is a bright display on any of the three LED banks. Now each of the individual frequency sensitivity controls can be varied to give the most dynamic display. The maximum sensitivity in each case is when the controls are fully anti-clockwise. For the display to be effective, the music input should cover all frequencies. After the three controls are set, it should only be necessary to vary the input sensitivity control to set the display. Although not supplied in the Dick Smith Kit of parts, a conventional pot with shaft and knob could be substituted for RV1 for a more convenient control, as described previously.

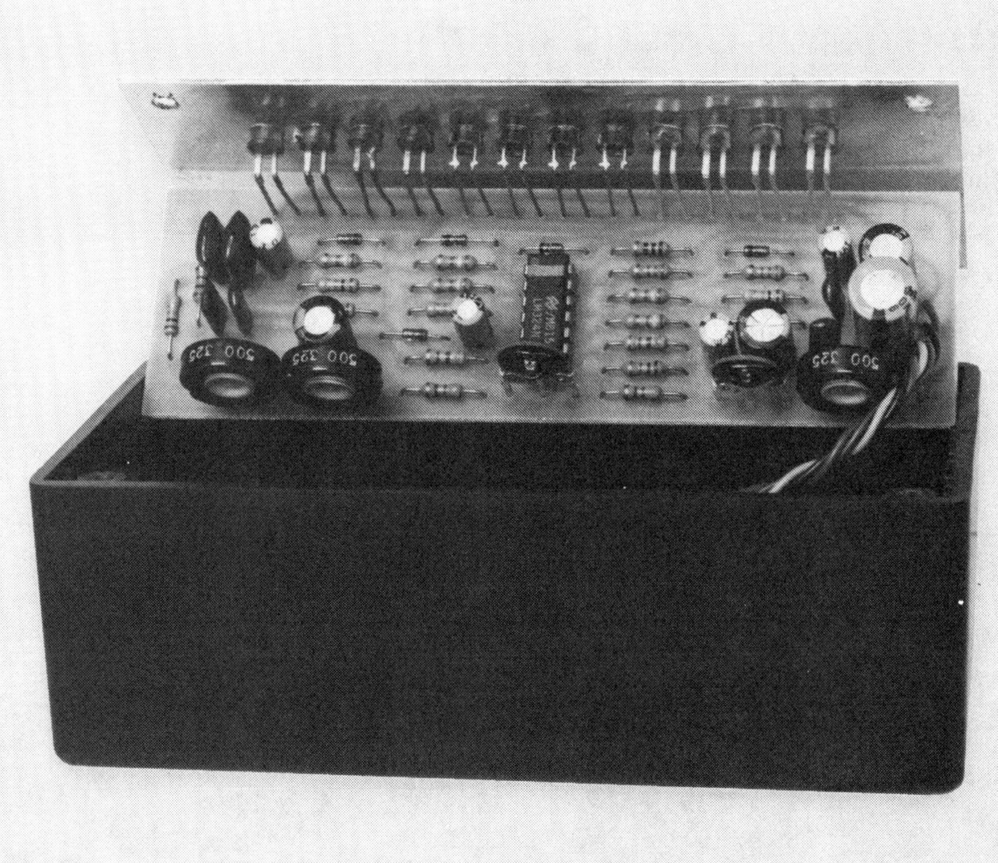

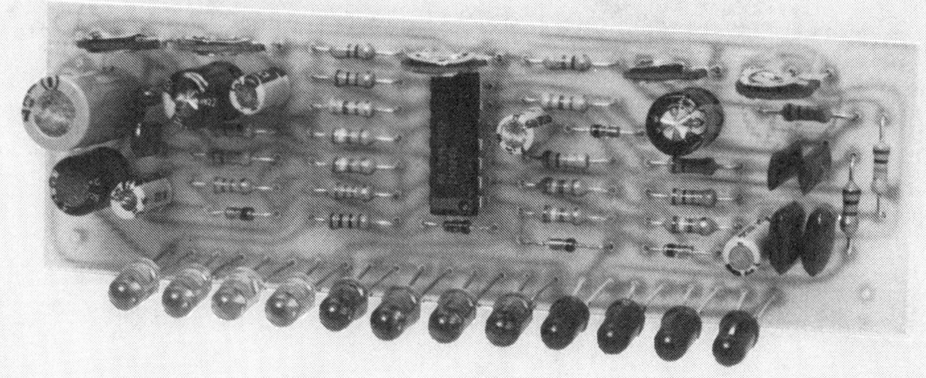

Top photograph shows the assembly in the optional Zippy Box.

Lower photograph shows the completed circuit board.

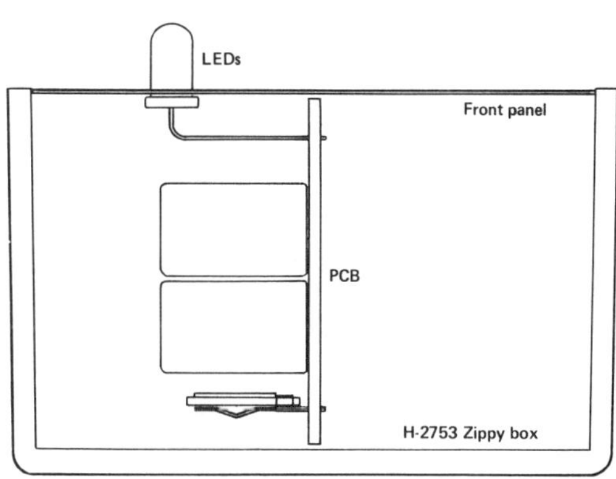

Right: Illustration shows the LED position relative to the front panel and PC board.

Combination Time Lock Switch

Based on CMOS IC's, this project is an excellent teaching/learning aid to logic design. Not only is it educational, but it can be the basis of a party game or even put to more practical serious use as part of an alarm/deterent circuit. To 'open' the lock, the correct combination of switch action is required, and this must be done within a specific time period. Incorrect combinations will result in an 'alarm' state. The output stages can be wired in different ways to suit different requirements.

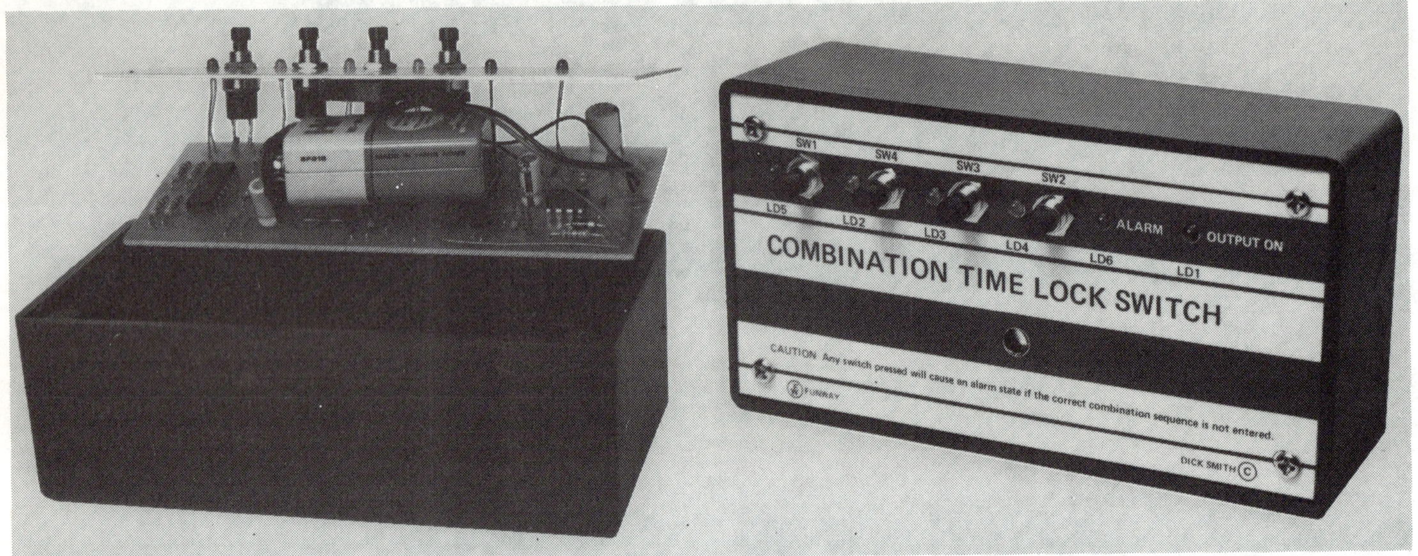

The Frustrator Party Game assembled ready to insert into the optional Zippy Box. Note how the 9 volt battery is located.

The completed unit shown in the optional Zippy Box.

PARTS LIST

Part Description		Total Quantity
Resistors:		
150 ohm	R22	1
1.5k	R6, R8, R10, R12, R15, R17	6
10k	R1, R2, R3, R4, R19	5
68k	R9	1
470k	R5, R7, R20, R21	4
1 Meg	R11, R13, R14, R16, R18 R23	6
Capacitors:		
220pF	Ceramic C9	1
.01uF	Ceramic C10	1
4.7uF	RB Vertical Electrolytic C1, C2, C4, C5	4
33uF	RB Vertical Electrolytic C3, C6, C8	3
100uF	RB Vertical Electrolytic C7	1

Part Description		Total Quantity
Semiconductors:		
4002	CMOS IC1	1
4013	CMOS IC2, IC3, IC4	3
555	Timer IC IC5	1
BC 337 or 2N2222 NPN Transistor TR1		1
1N4148 Silicon Diodes D1, D2, D3, D4, D5		5
1N4002 Power Diode D6		1
3mm Red LED's LD1, LD5, LD6		3
3mm Green LED's LD2, LD3, LD4		3
Miscellaneous:		
SP Pushbutton Switch SW1, SW2, SW3, SW4		4

Parts Description		Total Quantity
Piezo Transducer L-7022		1
Battery Snap P-6216		1
PC Board ZA-1466		1
Tinned Copper Wire 100mm		1
Hookup Wire – 1m.		1
Optional Components: (Not supplied in basic kit).		
Zippy Box Case UB 3 H-2753		1
12V. Coil Relay S-7120		1 or 2
Hookup Wire – assorted colours W-4010		1
9 volt Battery or Power Supply		1
Basic Kit of Components. Cat K-2666		

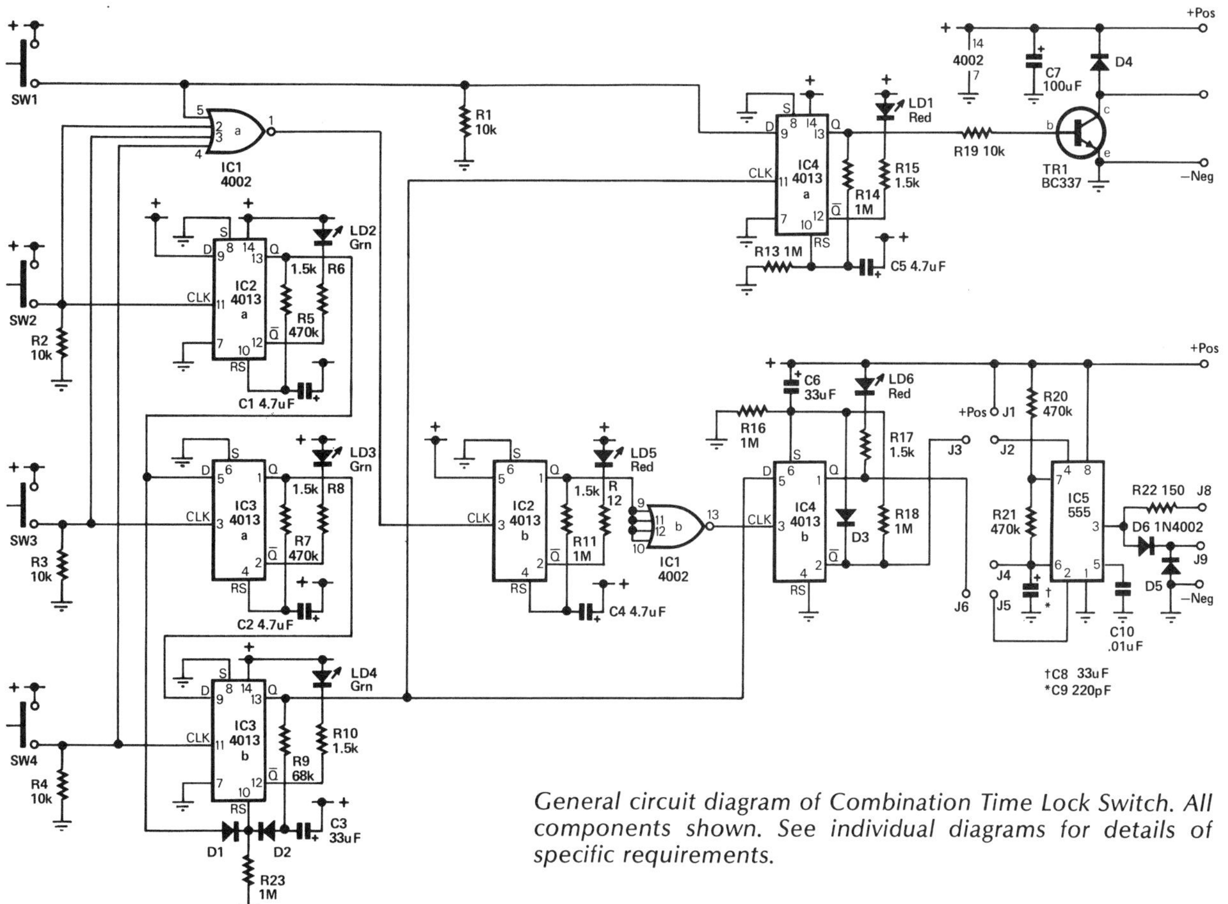

General circuit diagram of Combination Time Lock Switch. All components shown. See individual diagrams for details of specific requirements.

HOW IT WORKS

The basic operation of the unit requires that the operator has to press in sequence, four pushbuttons in order to switch the output stage. These switches not only have to be pushed in the correct order, but also, the action has to be carried out within a specific time period. A switch that is pressed out of turn or out of the correct time slot will result in an alarm state.

Each switch is connected to a simple monostable in each case, one half of a 4013 CMOS pack. These 'D' type flip-flops are configured as monos by connecting the 'Q' outputs back to the reset input via a resistor/capacitor combination. A LED (light emitting diode) is wired to the 'Q̄' output to show the mono in the active state.

A switch register IC2b, or more accurately, a switch released register is provided and is indicated by LED, LD5. A NOR gate, section (a) of a 4 input 4002 is used to provide the necessary gating to enable this function. If the correct sequence of

events is not entered, the alarm register, IC4b will be triggered via pin 3, the clock input, and the second half of the 4002 NOR gate configured as an inverter. The LED, LD6 will show this state.

To Switch The Output Stage

The sequence of events is initiated by pushing SW2. This applies a '1' to the clock input, pin 3 IC2a that activates this mono on the rising edge of the transition. This 'D' type flip-flop transfers the '1' on the data input, pin 9, to the 'Q' output, pin 13. The complementry output 'Q̄' goes low and consequently lights LD2. The 4.7uF capacitor C1 is now charged via the 470k, R5. When the voltage on pin 10 the reset input, reaches the threshold point, the flip-flop reverts to its original state. This mono period takes around 1.5 seconds.

It is within this period that SW3 has to be pressed to follow the correct series of events. The same action as

described above is repeated for this mono IC3a except in this case, the 'D' input, pin 5 receives '1' from the previous mono 'Q' output. This second action in the sequence is reliant on the first mono to be in the active state before it can be enabled. Again, the time out period for this mono is around 1.5 seconds.

While in this active state, SW1 has to be pressed, then slightly after, SW4. It is here that another parameter is entered to further increase the combination security. This third mono IC3b will not trigger unless the first is off and the second is on. If for example, IC2a is still on, 'Q' output, pin 13, will be high and therefore pin 10, the reset input of IC3b, will also be held high via D1. This holds the mono in the original state and any clock pulse from SW4 will be ignored.

Now assume that SW4 was pressed at the correct time, the mono will flip and the 'Q' output at pin 13 will go high and remain on for around 4 seconds. It is the positive going

HOW IT WORKS (cont.)

transition of this output that now triggers the control flip-flop, IC4a via the clock input, pin 11. SW1 has to be on at this time so that '1' is present on the data input, pin 9. This is transferred to the 'Q' output, pin 13. The transistor, TR1 is now switched on and therefore the load connected to its collector.

Depending on how the control flip-flop is configured, it will either reset after a period of around 4 seconds or remain on. The two resistors R13 and R14 determine this state. If wired with R14 left off the board, the circuit behaves as a flip-flop and has to be turned off. This is done by repeating the normal switch sequence except that SW4 is not pressed therefore transfering a '0' to the output and turning off TR1. If R14 if left on the board and R13 is removed, the circuit will behave as a monostable and automatically reset after a short period.

False States That Cause an Alarm

Unlike the switch registers, the alarm register output, pin 1 is high and normally set. As previously described, any switch pressed will activate the switch register, IC2b via the NOR gate. The time out period for this mono is around 3 to 4 seconds. If it is in the active state, the 'Q' output, pin 1 will be high. This is inverted by the NOR gate to make the clock input, pin 3 of the alarm register IC4b, low. If the correct sequence has not been entered, the data input, pin 5 of this register will be low because of the state of SW4 switch register, IC3b 'Q' pin 13. At the end of the IC2b time out period, the output will again revert to the original state and therefore the clock input at pin 3 of the alarm register will go high. It is on this positive going transition that the data on pin 5, (0) will be transfered to the output, pin 1. This then is the alarm state indicated by LD6 and the alarm output stage. This 555 timer output stage can be configured to operate in different ways to drive piezo transducers, speakers, relays etc. The alarm register itself can also be made to operate in different modes, either as a mono with a 1.5 second timeout or a flip-flop that remains on till such time as the normal correct turn off sequence is activated.

Other Circuit Detail

The alarm output stage 555 can be made to operate in an astable mode to be used as a siren. The output at J8 can be used to drive a piezo transducer or speaker. In this case it is switched on and off by using the reset facility at pin 4 linked through J2, J3. The timing capacitor, C9 220pF determines the output frequency. By varying the value of this capacitor, the frequency can be changed.

It can also be used as a monostable to give an output period determined by the timing capacitor connected to pin 6. With a 33uF capacitor, this period will be around 50 to 60 seconds. This mode uses the trigger input, pin 2 via links J5, J6 to control the switching state.

The control output stage uses a BC337 transistor to drive a relay. The contacts of this relay can be wired to the device that is to be switched on and off. The diode D4 is a transient suppressor to remove any inductive spike from the relay coil.

All mono timing capacitors are refered to the positive supply rail so that the correct state is set during the power up sequence.

Variation of Timeout Periods

The period of each mono can be varied by changing the timing resistor in each case. These are R5, R7, R9, R11, R14 and R18. If it is found that one or more of the timing periods are outside the range needed for normal operation, it may be necessary to change one or more of the above values to fit within the combination sequence.

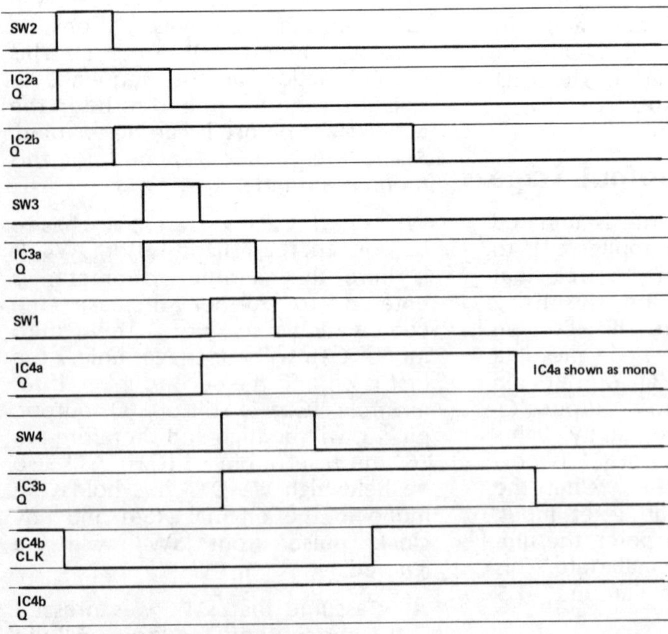

Typical normal switching sequence state timing diagram to turn output on.

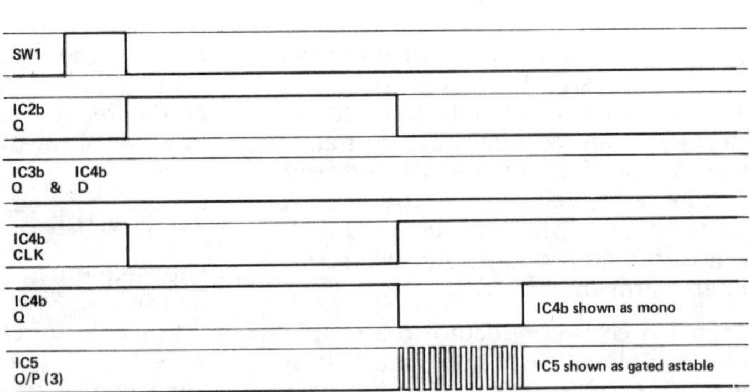

False state switching sequence timing diagram that results in an alarm state.

GENERAL ASSEMBLY

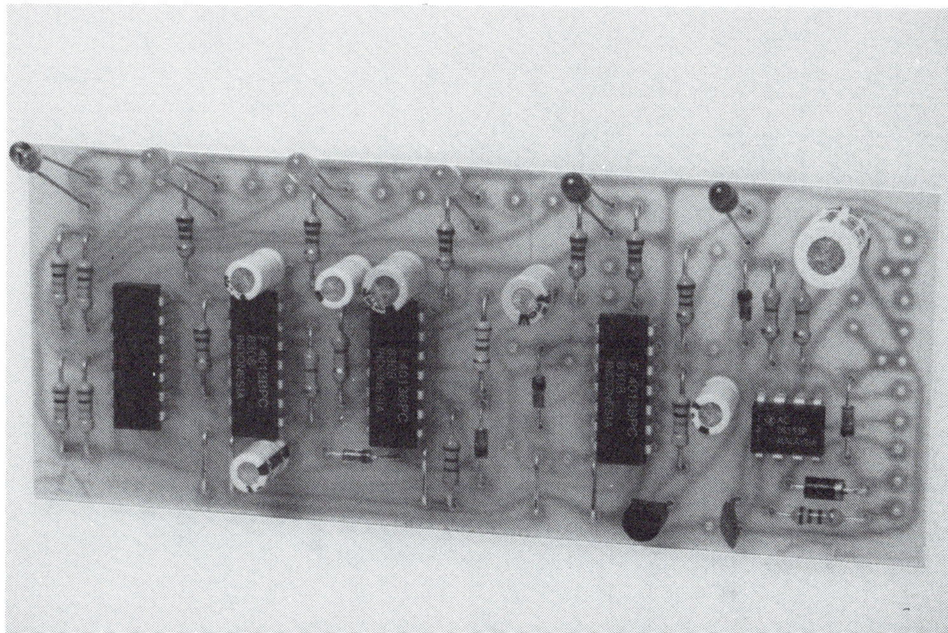

PC Board Assembly shown with common components to all variations of the basic circuit.

1. Insert all low profile components, the links L1 to L5, the diodes D1 to D6 and the resistors. Leave out R13, R14, R16 and R18 at this stage. Cut all pigtails and solder as required.

2. Now fit all capacitors, note the polarity of electrolytics. Leave out C8/C9 at this stage. Cut all pigtails and solder as necessary.

3. Insert ICs and transistor noting the orientation, solder in position.

4. The LEDS (light emitting diodes) and the pushbutton wiring will depend on how you wish to mount or house the PC board. See the section under Frustrator Assembly for how these components are wired and inserted.

5. The power, transducer, speaker or relay wiring can be done in accordance with the requirements of the final use.

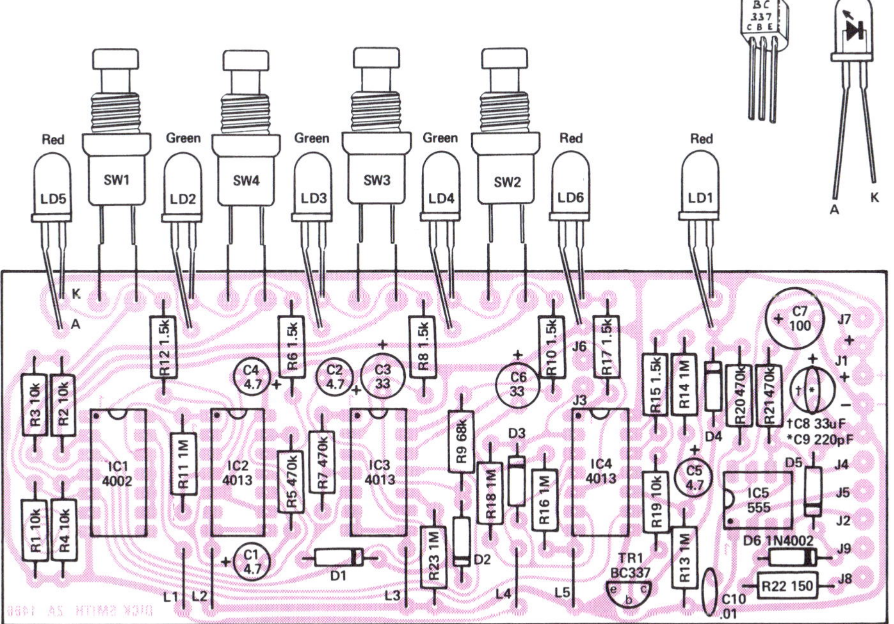

Combination Time Lock Switch General PC Board Overlay Diagram. All component positions shown.

Frustrator (Party Game)

The object of this game is to get the player to 'break' the combination of the switch and turn on the control LED, LD6. Without prior knowledge of how the unit works or how the sequencing is to be entered, there is little chance of this happening. Only when vital clues are given is there some chance of opening the lock. For example, the switch sequence could be given as the first clue, the time out periods could be the second.

Every time the wrong state is given, the alarm LED, LD6 comes on and the piezo tranducer or speaker (if fitted) will sound. This state could be used as player disqualification.

For those friends who may have already had the advantage of reading this book, they can be fooled by wiring the switches, and the LEDS if you wish, in a different way to that described. For example, if SW1 is wired to where SW3 is normally connected and vice versa, a completely new combination is coded. You can see that many combinations can be made up simply by changing the wiring in this way.

Sobriety Tester

The reaction time of a person can be tested at an adult party with the Frustrator. A 'player' that is known to be competent in using the combination switch in a sober state, may not be so capable after a few drinks. Observers may be surprised at the reactions of such persons who considers themselves sober. It may be that more than one attempt is required before the combination is cracked.

Assembly

For this game and the sake of convenience and handling, we have arranged the board and wiring to fit into a UB3, H-2753 Zippy Box.

Firstly, the alarm register, the alarm output stage and the control register will have to be wired to suit this application.

R18 is inserted into circuit, R16 is left out of the alarm register. This makes it a short monostable to light LED, LD6 and trigger the 555 output stage. Output links, J3 and J2 are joined.

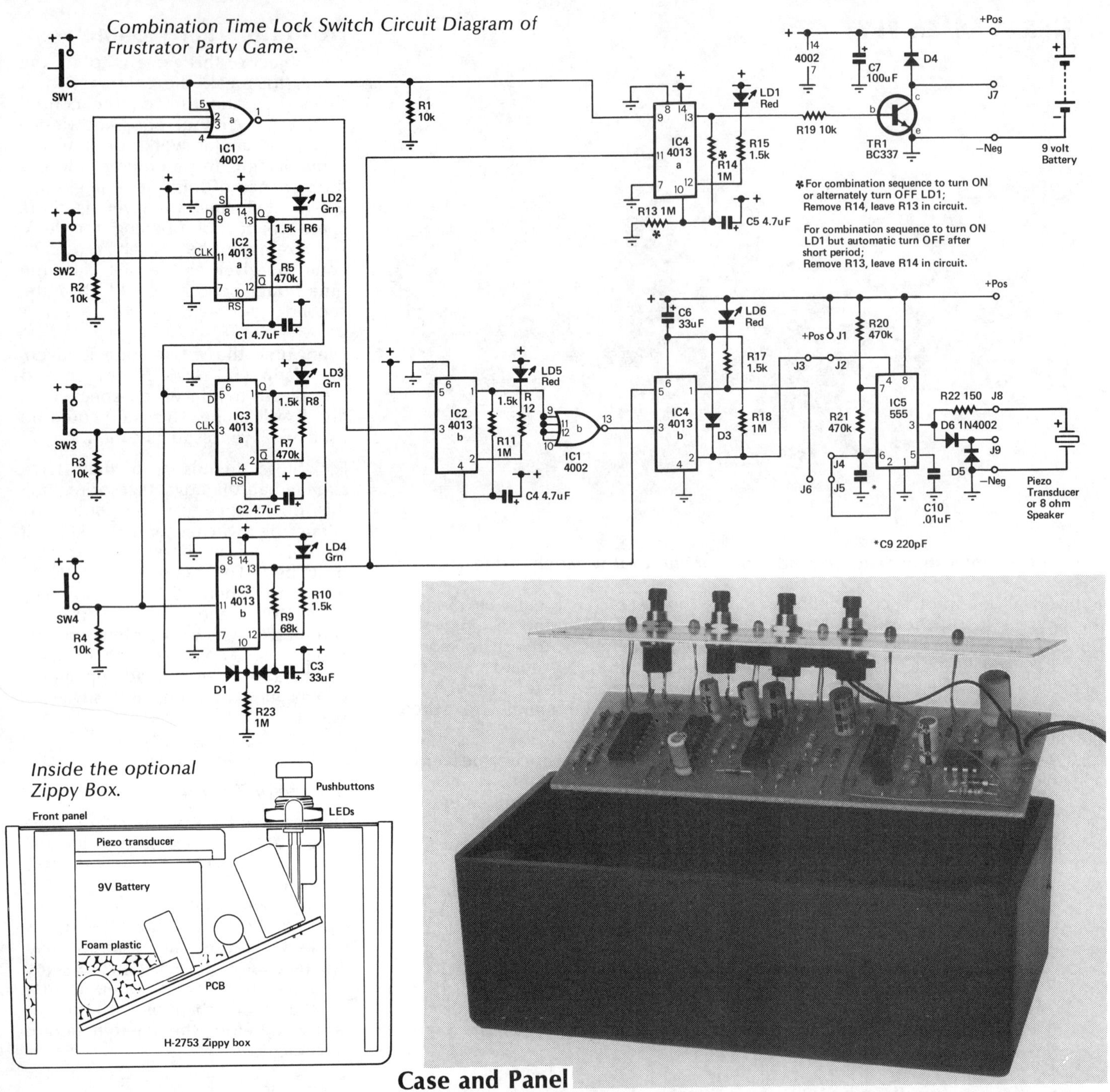

The alarm output stage is wired as an astable by including a 220pF capacitor in the C9 position, J4 is joined to J5. The piezo transducer is wired to the negative power rail and J8 as shown, this produces the alarm sound.

The control register is configured to operate as a normal flip-plop by inserting R13 into circuit; R14 is left out. Note that the transistor TR1 is not used in this application.

The power source used for this project is a 9 volt battery and the wire and snap connector have to be connected as shown.

Case and Panel

Neatly cut out the front panel from the rear of this book. Place it accurately over the Zippy Box lid and mark the LED, switch and transducer hole centres at the points shown. Drill carefully 6 x 3mm holes for the LEDS, 4 x 7mm holes for the switches and 1 x 5mm hole for transducer. With a suitable rubber based contact adhesive, stick the label accurately onto the lid so that it aligns at the edges. To punch out the holes in this label, use the rear end of the drill shank to push through the hole from the front face to the rear. Use the 3mm drill to punch out the case screw holes.

The four pushbutton switches can now be fitted to the front panel as shown. Turn the body of these switches so that the terminals align with the holes in the circuit board. Now wire short lengths of tinned copper wire to each terminal (these can be the offcut pigtails from resistors). The PC board can now be threaded onto these wires and pushed up to about 2mm of the terminals. Hold in this position and solder all leads, cut off excess. Now bend the board away from the panel as shown till it sits at an angle of about 15 degrees from the front panel. Try fitting the board and panel

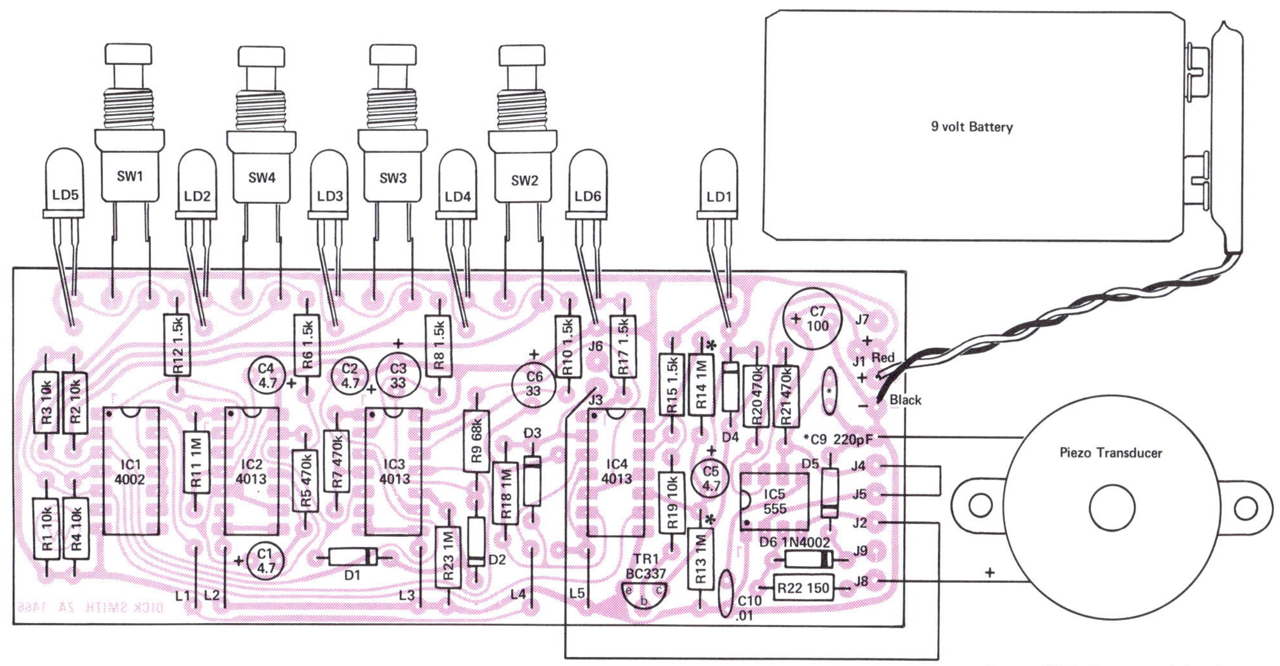

Frustrator Party Game PC Board Overlay Diagram.

✱ For alternate turn ON, turn OFF action;
Remove R14, leave R13 in circuit.

For turn ON via the correct switch action
but with automatic turn OFF;
Remove R13, leave R14 in circuit.

into the case. It should be adjusted so that the assembly fits snugly in between the mounting posts inside the case.

To fit the LEDS, simply thread the legs (observe the polarity) through the appropriate holes in the board, adjust and bend slightly so that the body penetrates and fits squarely into the panel hole. Solder into position.

The piezo transducer is simply glued to the rear of the panel over the hole previously drilled. The 9 volt battery is located over IC3 and IC4 bodies, between C1 and TR1. A small piece of foam can be used to hold it in place. The whole assembly can now

be inserted into the case, it should fit snuggly into position.

As you can see, we have not provided any on-off switch with this circuit. The object of the game would be somewhat lost if the alarm could be stopped by turning off the power switch. For those who wish to include such a facility, this could be done by connecting a small switch into the positive lead of the battery lead and mounting it somewhere in the case. As the circuit only consumes a small amount of current in the standby state, around 5mA at 9 volt, the screws could be tightened and the unit left on for the life of the party.

Variations

The switches could be wired in other ways to change the combination. This would involve using a small amount of hookup wire to carry out the modification.

The alarm register could be modified so that the alarm sound stays on rather than the short period as described. This is implemented simply by removing R18 and inserting R16. Now the alarm has to be turned off by the correct sequence. This mode could be used as an embarassment piece for those people who insist on fiddling with things. Great for loungeroom tables, offices, waiting rooms, board rooms etc.

COMBINATION TIME LOCK SWITCH WITH RELAY CONTROL OUTPUT.

With this application, the control output is put to use to drive a relay. The contacts of this relay could be considered simply as a small switch and used as a substitute for an on-off switch in an alarm system. You may wish to have a small alarm on your bedroom, locker, toolbox, workshop, darkroom, glasshouse, car or bike. The combination lock switch in this circumstance would have to be interrogated before entry or access is gained into this area. The red LED, LD1 indicates if the switch is on or off. Any attempt made to 'fiddle' the combination will result in a short alarm signal from the transducer or speaker, whatever used.

Circuit Detail Changes

Control Register

The control register IC4a is configured as a flip-flop in this application and is wired as shown. R13 is wired in circuit, R14 is left out. This circuit then behaves as an on-off switch controlled by the correct combination sequence. The transistor TR1 drives a small relay. Any 12 volt coil type with a resistance of 120 to 350 ohms could be used. The type shown is only suitable for low voltage, low current switching applications.

Alarm Register

The alarm register, IC4b is wired as a short monostable in this case; R18 is left in circuit, R16 is left out. The 555 output stage is switched on and off by this mono at the reset input, pin 4 via J3, J2. The output is connected to a piezo transducer as the warning device. A speaker horn could be used as a substitute but it must be realized that the output from this system is not great.

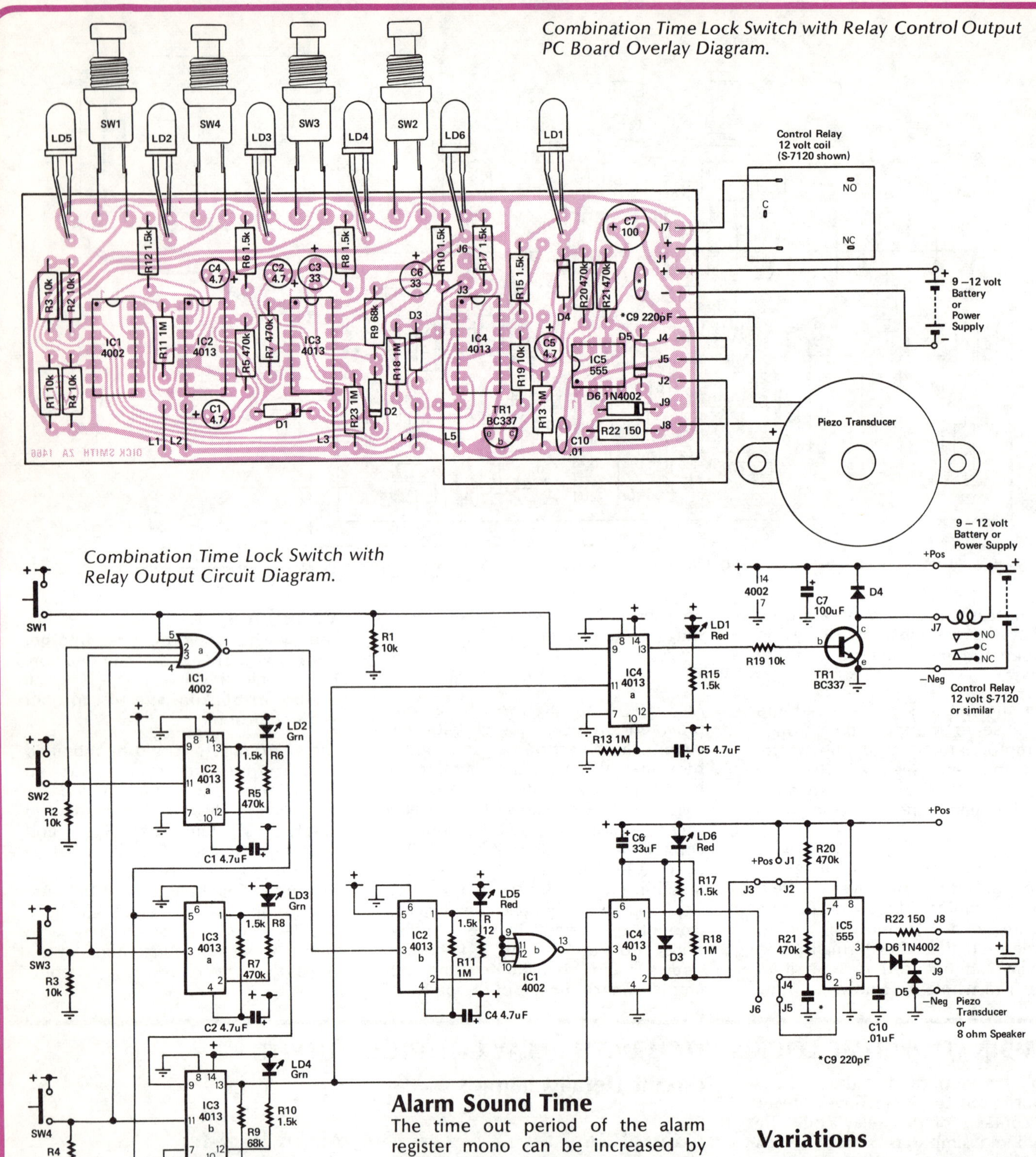

Alarm Sound Time

The time out period of the alarm register mono can be increased by making C6 a larger value. The physical size of this electrolytic will tend to limit what can be used in the board.

Vehicle Disable

This simple combination lock can be applied to a vehicle to prevent the motor being started. By connecting the relay contacts across the points of the ignition system, when the combination is on, the points are shorted and any attempt to start the motor will result in failure.

Variations

This basic function is also applicable to a drunk, don't drive situation. Because the combination sequence requires a specific time control, a person who has had too much alcohol to drink will not be able to interpret the switch time intervals correctly and end up not being able to start the vehicle.

This application should not be taken too seriously but is a thought that can be experimented with under controlled supervision.

WARNING

The system could be used with relays with a 12 volt coil and 240 volt mains rated contacts. In the interests of safety, we strongly recommend that an application used to switch mains operated equipment is left to the experts. Experimenters should not enter into this area.

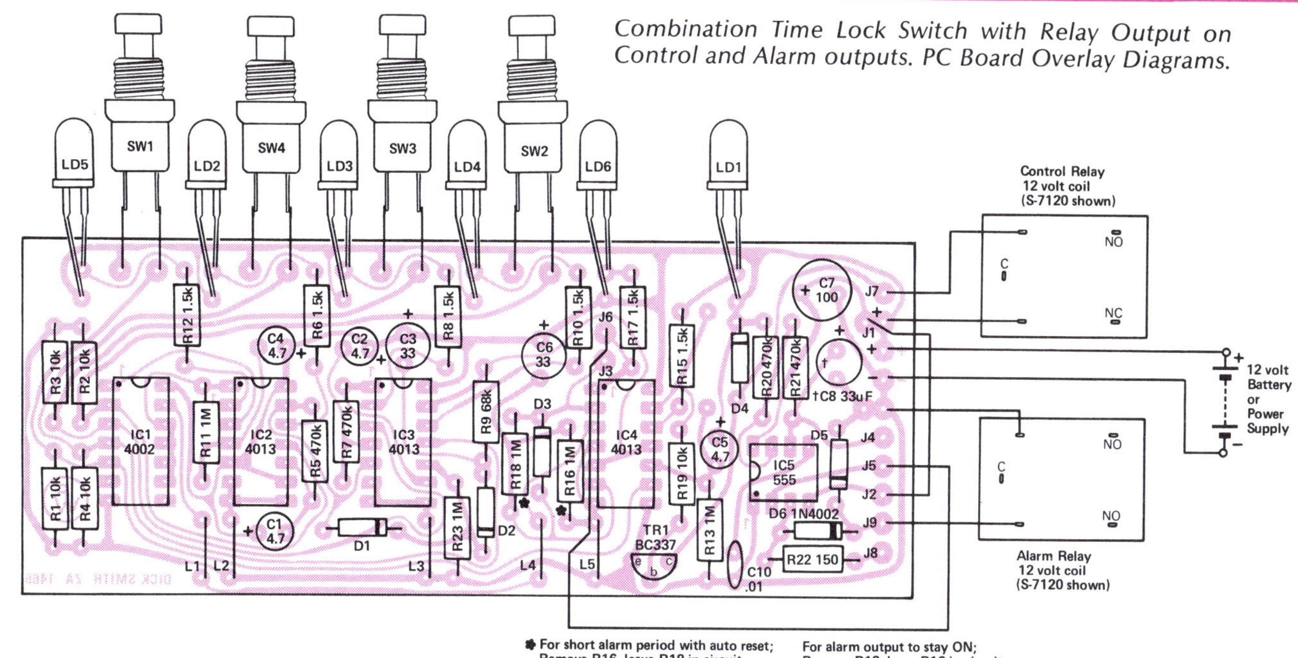

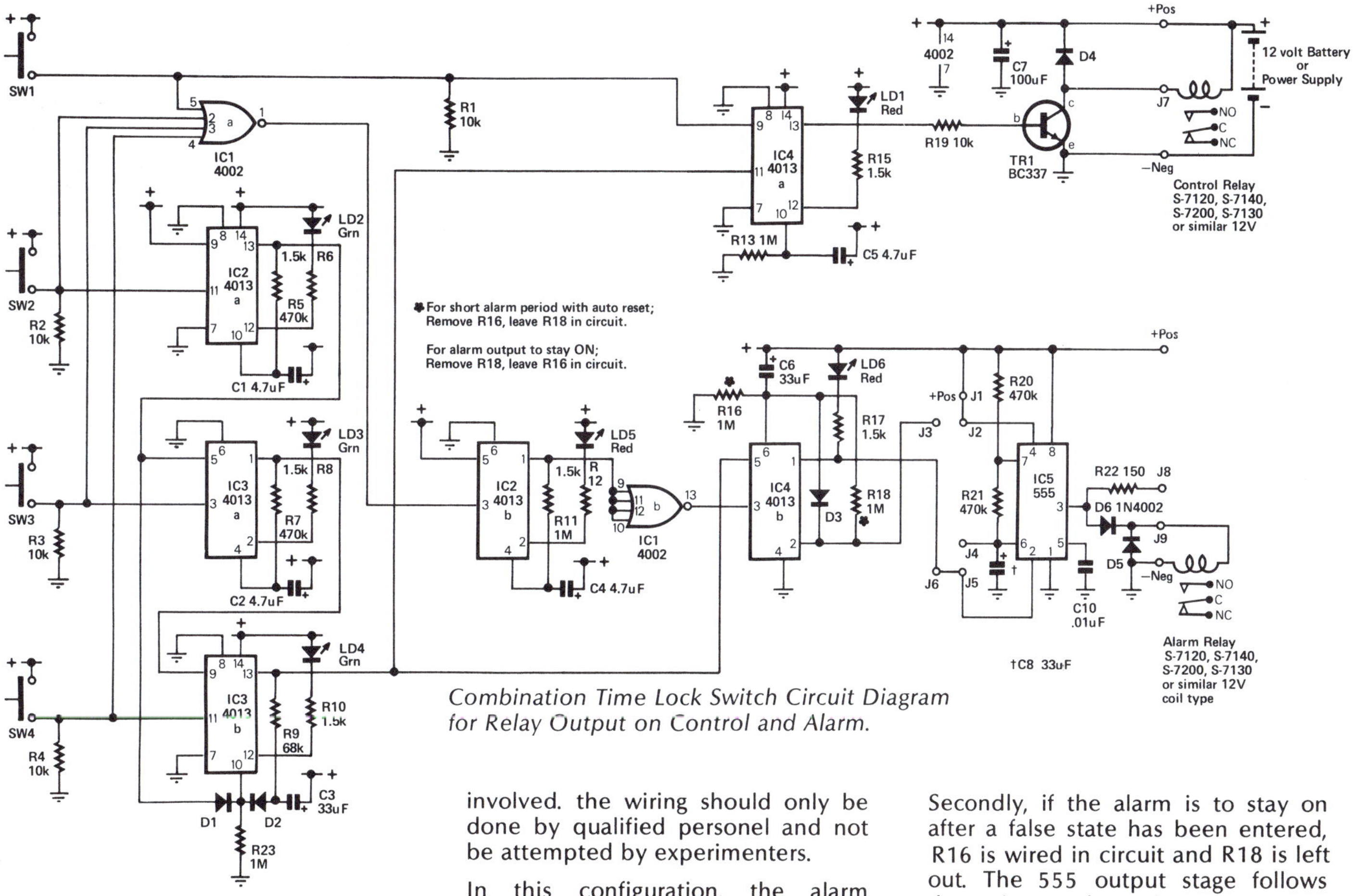

Combination Time Lock Switch with Relay Output on Control and Alarm outputs. PC Board Overlay Diagrams.

Combination Time Lock Switch Circuit Diagram for Relay Output on Control and Alarm.

Alarm Relay Output

A relay can be connected to the 555 alarm output stage to give a more flexible use of this function. This relay could be used to switch sirens, bells, alarms, lights etc. Again, if 240 volt mains equipment is to be involved. the wiring should only be done by qualified personel and not be attempted by experimenters.

In this configuration, the alarm register can be wired in two ways.

Firstly, if a short alarm period with auto reset is required, R18 is wired in circuit, R16 is left out. This produces a short on period to trigger the 555 that is also configured as a mono with a time out period of around 30 seconds.

Secondly, if the alarm is to stay on after a false state has been entered, R16 is wired in circuit and R18 is left out. The 555 output stage follows the register and remains on.

In both cases, the output stage is wired with J1 joined to J2, J6 is joined to J5. The relay is connected between J9 and the negative supply rail. The diodes, D5 and D6 are required for correct operation of the 555 in this relay drive mode.

Li'l Pokey

One of the more complex projects in this series, yet simple to construct and get operational. Again based on CMOS devices, this project will be a barrel of fun for all you free wheelers or dealers. The display is based on LED devices and the self contained circuit operates from batteries. If you can resist the urge to play this great little game of chance before it is completed, it will provide many hours of fun for all ages. You can bet your bottom dollar that this project will go down well at parties.

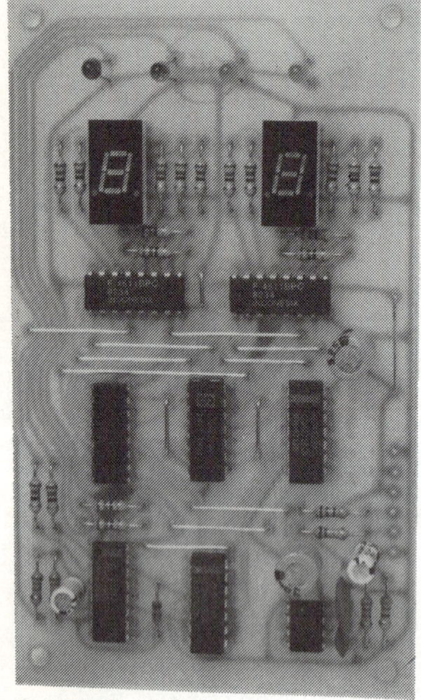

The completed circuit board prior to final assembly.

The project completed shown in the optional Zippy Box.

PARTS LIST

Parts Description		Total Quantity
Resistors:		
150 ohm	R18	1
680 ohm	R1 – R14, R21 – R24	18
33k	R16, R20	2
330k	R19	1
560k	R15	1
4.7 Meg	R17	1
Capacitors:		
.01uF	Ceramic C4	1
.047uF	Ceramic C3	1
4.7uf	RB Vertical Electrolytic C2, C5	2
100uF	RB Vertical Electrolytic C1, C6	2
Semiconductors:		
4001	CMOS IC6	1
4009	CMOS IC8	1
4030	CMOS IC5	1

Parts Description		Total Quantity
4071	CMOS IC7	1
4511	CMOS IC1, IC2	2
4518	CMOS IC3	1
555	Timer IC IC4	1
LT303/LT313	7 Seg. Display DIS1, DIS2	2
3mm Green LED	LD1, LD3	2
3mm Red LED	LD2, LD4	2
1N4148	Silicon Diode D1	1
Miscellaneous:		
Miniature DPDT Slide Switch	SW2	1
Miniature SP Pushbutton	SW1	1
Battery Snap	P-6216	1
4 x AA Battery Carrier	P-6114	1

Parts Description		Total Quantity
Tinned Copper Wire	500mm	1
Hookup Wire	200mm	1
PC Board	ZA-1462	1
Optional Components: **(Not supplied in basic kit).**		
Zippy Box Case UB1	H-2751	1
Piezo Transducer	L-7022	1
25mm Hex. Threaded Spacers	H-1847	4
4BA x 12mm Screws	H-1082	8
4BA Hex. Nuts	H-1332	4
Piece of Foam Packing	Approx. 150 x 70 x 20mm	1
Piece of Red Transparent Filter	55 x 25mm	1
Rubber Feet	H-1745	1
Basic Kit of Components **Cat K-2662**		

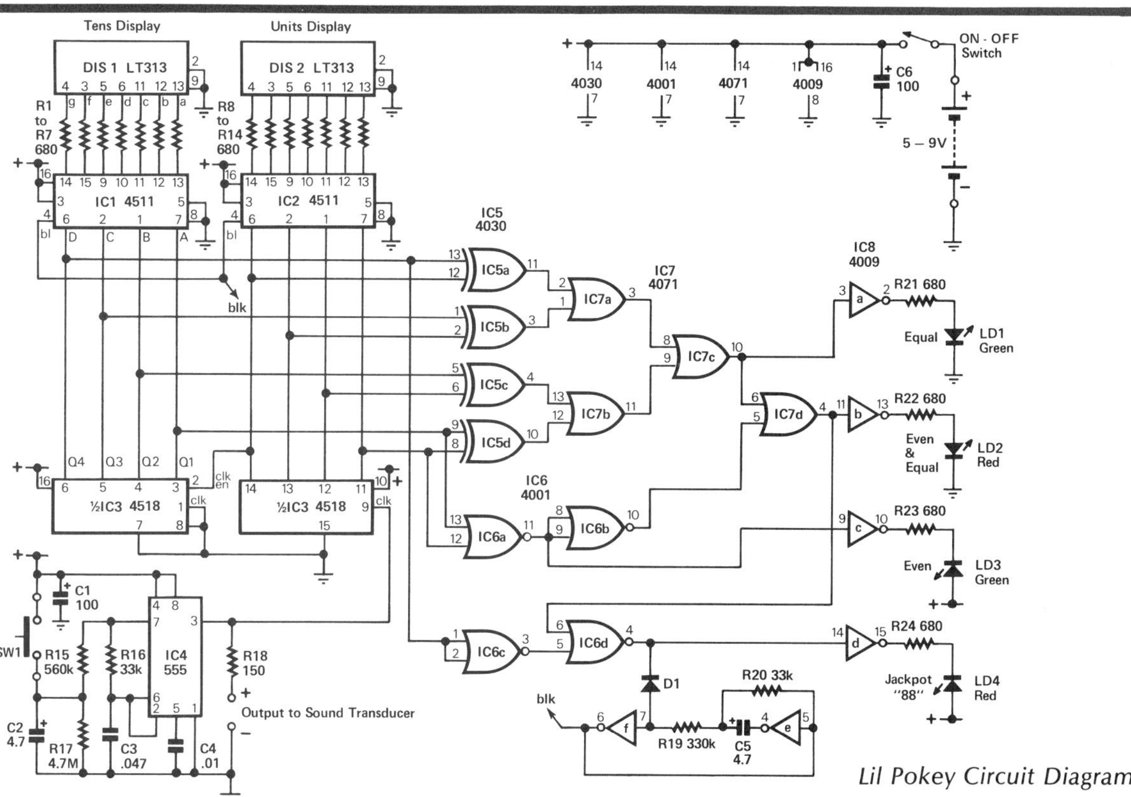

Lil Pokey Circuit Diagram

HOW IT WORKS

The circuitry consists of five main sections. The display uses two LT313 seven segment LED readout devices driven by two 4511 CMOS decoder/drivers. Two decade counters precede this display, both in one 4518 CMOS pack. The clock pulse to these counters is supplied from a 555 timer configured as a voltage controlled oscillator initiated by the pushbutton SW1. A decoding system made up of 4030, 4001 and 4071 gates register a win state at various levels. Win indicators in the form of single LEDS, LD1 to LD4 are driven by 4009 inverter/buffers.

The display section is straight forward and uses common cathode seven segment readouts. The 4511 BCD to seven segment decoder/drivers perform the necessary current sinking to the displays via the 680 ohm limiting resistors. It also provides the decoding between the counters and the seven segment format.

The clock pulse is derived from a 555. The circuit used needs a little explanation. The circuit is configured basically in an astable mode with one exception. The resistor (R15) that would normally be connected to pin 7 and taken to the positive rail to act as the charge rate component, is in this case taken to a storage capacitor C2. When SW1 is pressed, the voltage on C2 is raised to the full positive rail potential. This would then represent the normal astable mode of operation. When the switch is released, C2 is discharged mainly through R15 into the discharge, threshold and trigger input of the 555, pins 7, 6 and 2 respectively. As the voltage decreases, the charging current through R15 into the timing capacitor C3 will be reduced proportionally and therefore the charge period will be longer. This of course slows down the clock frequency and reaches a point where it eventually stops. This happens when the voltage on the threshold sensing input pin 6 falls below two thirds of the rail voltage. Resistor R17 is included to assure complete discharge of this control system. This circuit action represents the pull of the handle and the rolling action of the wheels. The output at pin 3 of the 555 is also used to drive a piezo transducer to simulate (very roughly) the 'rolling noise' of the machine cylinders.

The output of the counters is taken to a decoding network to register the various numbers that are interpreted as win states. The quad EXCLUSIVE OR gate, along with the three input OR gates, decode an equal state when both digits of the counters are the same. The state is buffered by the 4009a inverter and registered by the green LED, LD1.

Gate IC6a of the quad two input 4001 NOR pack decodes the least significant bit of each counters BCD output code to register an even number via inverter 4009c and the green LED, LD3. The OR gate IC7d uses the decoded equal signal at IC7c together with the even signal, via inverter IC6b, to register an equal and even state at the red LED, LD2.

The 'jackpot 88' is decoded in a roundabout way by monitoring the most significant bit of the 'tens' counter BCD output code and the even and equal states. If a '1' is on this most significant bit, pin 6 of 4518, the decimal number for this 'tens' digit can be either 8 or 9. If at the same time, the decoder finds that the evens condition is true, then the number must be '8'. This satifies the 'tens' decade state but dosn't include the 'unit' decade. This is done simply by looking for an even state of both decades. In this way, the number must be '88'. Gates IC6c and d perform this function via the even and equal state at IC7d. The jackpot is registered at the red LED LD4.

A simple additive circuit surrounding the 4009 inverters e and f is used to emphasise the jackpot win. When a decoded '88' appears, the output at IC6d goes high and reverse biases the diode, D1. This enables the low frequency astable oscillator made up from these two inverters. The output is connected to the display driver blanking inputs. This causes the seven segment displays to flash on and off at the frequency of this oscillator thereby greatly emphasising the exciting fact that you've won!

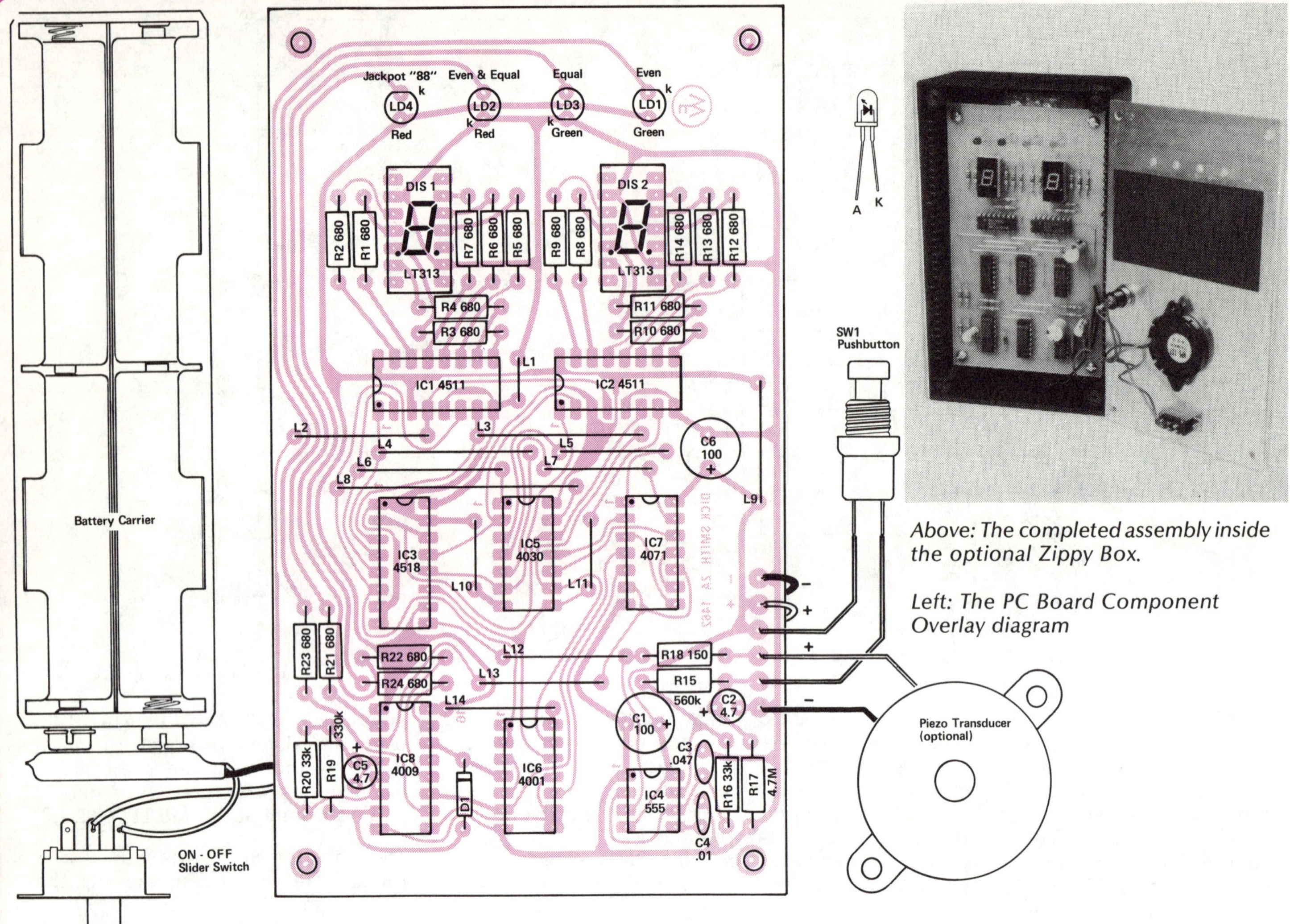

Above: The completed assembly inside the optional Zippy Box.

Left: The PC Board Component Overlay diagram

CONSTRUCTION

1. Check off all components supplied in the kit against the parts list.

2. Insert all the low profile components – firstly the links 1 to 14. These can be formed from tinned copper wire by cutting to the appropriate length and inserting in the positions shown. Make sure each length is straight and flat against the board. Your long nose pliers can be used to pull the lead through and shape the bends. Make the bends on the underside sharp and hard on the shoulders of the hole so that the link will remain solid in position while cutting and soldering. Next load the resistors and diode D1. Cut and solder in position.

3. Insert all capacitors, observe the polarity where necessary, cut and solder in position. All IC's can now be inserted and soldered. Do not include the displays at this stage. See section 5 following for detail of how these are inserted.

4. The housing arrangement has to be considered at this point. If you follow our example and use the H-2751 Zippy Box, it is necessary to mark and setup the PCB inside this case. It is held off the bottom of the box by four 28mm spacers. We used 4BA x 25mm threaded spacers and 4BA, 3mm thick hex nuts to achieve this standoff distance. This setup is shown in the diagram. Other mounting means could be used but this would be up to the individual constructor. To mark the mounting hole positions, simply place your PCB, copper side down, centred in the bottom of the box. Now using the board as a template, mark each position through the holes with a pencil or scriber. At these marked centres, drill 4 x 4mm holes.

5. The seven segment displays can now be mounted on the PCB. Depending on the type supplied, there may only be 10 legs on the pack, not 14 as drilled in the board. In this case, simply mount the display in the centre of the group; the two holes at each end are not used. Don't push the pins all the way through, only about 1mm need penetrate. This stands the display up off the board so that it is closer to the front panel window. Solder into position if this is to be the mounting method you wish to adopt. For those constructors who want the display as close to the front panel as possible, two optional 14 pin IC sockets can be soldered in each location. In this way, the socket lifts the display further off the board.

6. The LEDS can now be inserted. Observing the polarity, push through the leads untill the body of the LED is high enough to penetrate the front panel (see inside case diagram on the following page). This will require measurement to get the correct lead length. Don't cut the leads untill you are sure this is correct.

7. Wire the battery snap and on-off switch as shown. A 75mm length of wire is required to connect this switch to the board. The pushbutton switch can be connected and soldered with 2 x 50mm lengths. If an optional piezo transducer is to be added for sound effects, this can

now be wired. This completes the assembly and wiring of the PCB.

8. Now compare your assembly against the overlay diagram. Check that values of components are loaded into the correct position, note all polarity orientation. Take a close look at the underside of the board for bad soldering or tracks that may be shorted out by small slivers of solder.

Fit batteries to the carrier, push on the snap and switch on. Some form of display should result. Press the pushbutton and release. The display should run, fast at first then gradually slow down. After two or three seconds, it should stop. You will notice that the 'units' display (RHS), will run much faster than the 'tens'. This is the correct operation as the first counter divides the clock frequency by ten before reaching the second. Run through the operation several times to see if the win indicator LEDS operate as per the winning table. Obviously the 'evens' will come up on a fairly regular basis whereas the 'jackpot' may take all night.

9. Front Panel

After you have resisted the playing urge, the board and box assembly can be finished. The front panel label can be cut from the rear of this book. Position the label accurately over the Zippy Box lid so the outline aligns with the edges. With a sharp scriber, centre punch or similar instrument, carefully mark the centre for LEDS, transducer and switches. At the corner of the rectangular cutouts for the displays and slider switch, lightly and accurately mark these points. Remove the label. Now with a rule

or similar straight edge, join the dots you have just marked so that you end up with the outlines of the rectangles clearly visible. The 3mm holes for the LEDS and the transducer, the 7mm hole for the pushbutton and the 2.5mm holes for the slide switch can now be drilled. Be careful with this operation: don't let the drill slip or move off the hole centres. The rectangular holes can now be cut out. This can be done with a piercing, jewellers or coping saw and the edges dressed with a needle file or similar. If these types of saws are not available, a needle file or similar can be used to shape the hole. In both cases it will be necessary to drill a small pilot hole in the centre of the cutout to allow access of the saw blade or file. Again, be very careful with this procedure; it would be a pity to ruin the project by overshooting the display window outline or damaging the panel.

The label is now positioned over the panel after coating the rear with a rubber based contact adhesive. Make certain it is in the correct position before securing. Now with the shank of the drill previously used to drill the holes, punch out from the front surface to the rear, the appropriate LEDS, switch, transducer and mounting points. The rectangular sections for the switch and display window should be cut out accurately with a razor blade or similar instrument. If desired, a small piece of red filter material can be glued to the rear of the display window. The on-off switch, the piezo transducer and the pushbutton can be mounted on the panel.

10. Case Assembly

Mount the 4 x 4 BA screws and the hex nuts on the PCB and tighten securely.

Screw on the 25mm spacers and tighten.

The battery carrier is placed in the bottom of the Zippy Box and a piece of foam plastic is centred over the top. The PCB is fitted over the top and 4 x 4BA screws fitted through the bottom of the case used to hold the total assembly in position. When the screws are tightened, the battery carrier is held firmly in place by the foam. The piezo transducer (if fitted) can be glued to the rear of the panel over the previously drilled holes. The panel can be fitted to the box and the mounting screws secured. Four rubber feet can be fitted to the bottom of the box. This completes the assembly.

TO PLAY THE GAME

This is a game of chance and requires selected numbers to come up on the display to effect a winning combination. The table listed below shows these combinations that are indicated by the LEDs. The score weighting is shown at the right of this table. The winning states that are most likely to occur are two even digits and are rewarded with only a small score. The least likely to come up is the number 88. This number is considered the Jackpot and is emphasized by a flashing display.

How you use the score system is up to the players but we feel sure that many an entertaining hour can be spent. Lil Pokey will be the centre of attention at any party.

Win States	Score Indicators				Score
Both even digits	0	0	0	✳	2
Equal digits	0	0	✳	0	5
Even & equal digits	0	✳	✳	✳	10
Jackpot 88	✳	✳	✳	✳	50

The assembly shown inside the optional Zippy Box.

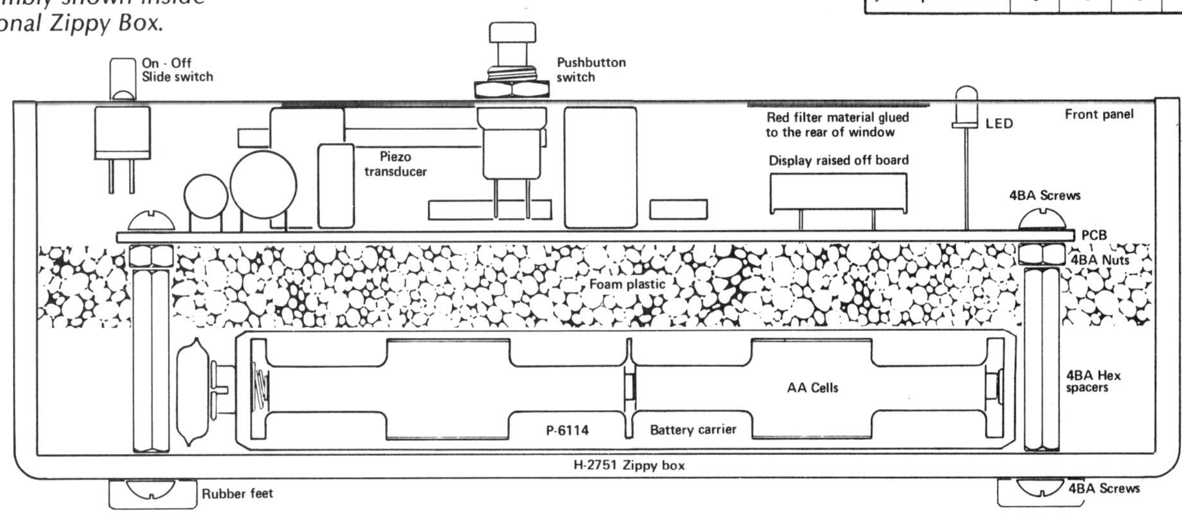

Binary Bingo

Logic design and display functions are the centre of this project. When completed, it becomes a game of skill and reaction time. Although apparently simple in playing requirements, it can be very frustrating and demanding. All that is required is to match the decimal number that comes up on the display with a series of four push buttons that represents a binary equivalent input. Sounds pretty simple, huh? Try it and say that again! The skill level can be changed by selection of a 'card count' switch.

The completed game shown here in the optional Zippy Box.

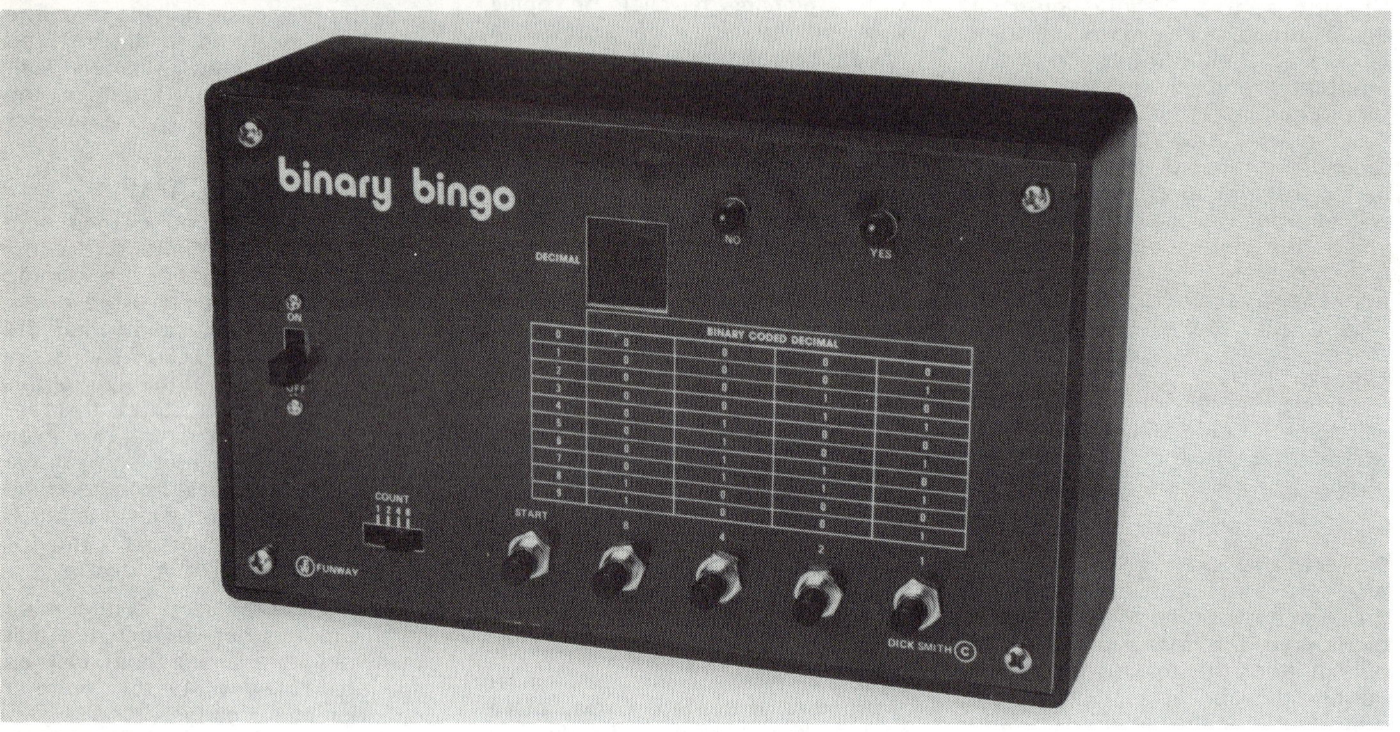

PARTS LIST

Parts Description		Total Quantity
Resistors:		
470 ohm	R6, R7, R8, R9, R10, R11, R12, R13, R16	9
2.2k	R17, R25	2
10k	R1, R2, R3, R4, R5, R14, R19, R20, R21, R24	10
56k	R23	1
120k	R22	1
1 Meg	R15, R18	2
Capacitors:		
.01uF	Ceramic C2	1
.022uF	Ceramic C1	1
10uF	RB Vertical Electrolytic C3, C4, C5	3
470uF	RB Vertical Electrolytic C6	1

Parts Description		Total Quantity
Semiconductors:		
4009	CMOS IC7	1
4013	CMOS IC4	1
4030	CMOS IC3	1
4511	CMOS IC1	1
4518	CMOS IC2	1
741	Op-Amp IC5	1
555	Timer IC IC6	1
LT303/LT313	7 Seg. Display DIS1	1
1N4148	Silicon Diodes D1, D2, D3, D4, D5	5
5mm Red LED	LD1	1
5mm Green LED	LD2	1

Parts Description		Total Quantity
Miscellaneous:		
SP Pushbutton Switch	SW1 – SW5	5
SP4P Miniature Slide Switch	SW6	1
DPDT Miniature Slide Switch	SW7	1
Battery Snap		1
4 x AA Battery Carrier		1
Tinned Copper Wire	300mm	1
Hookup Wire	150mm	1
PC Board	ZA-1468	1
Optional Components: (Not included in basic kit).		
Zippy Box Case UB1	H-2751	1
AA Cells		4
Basic Kit of Components. Cat K-2668		

HOW IT WORKS

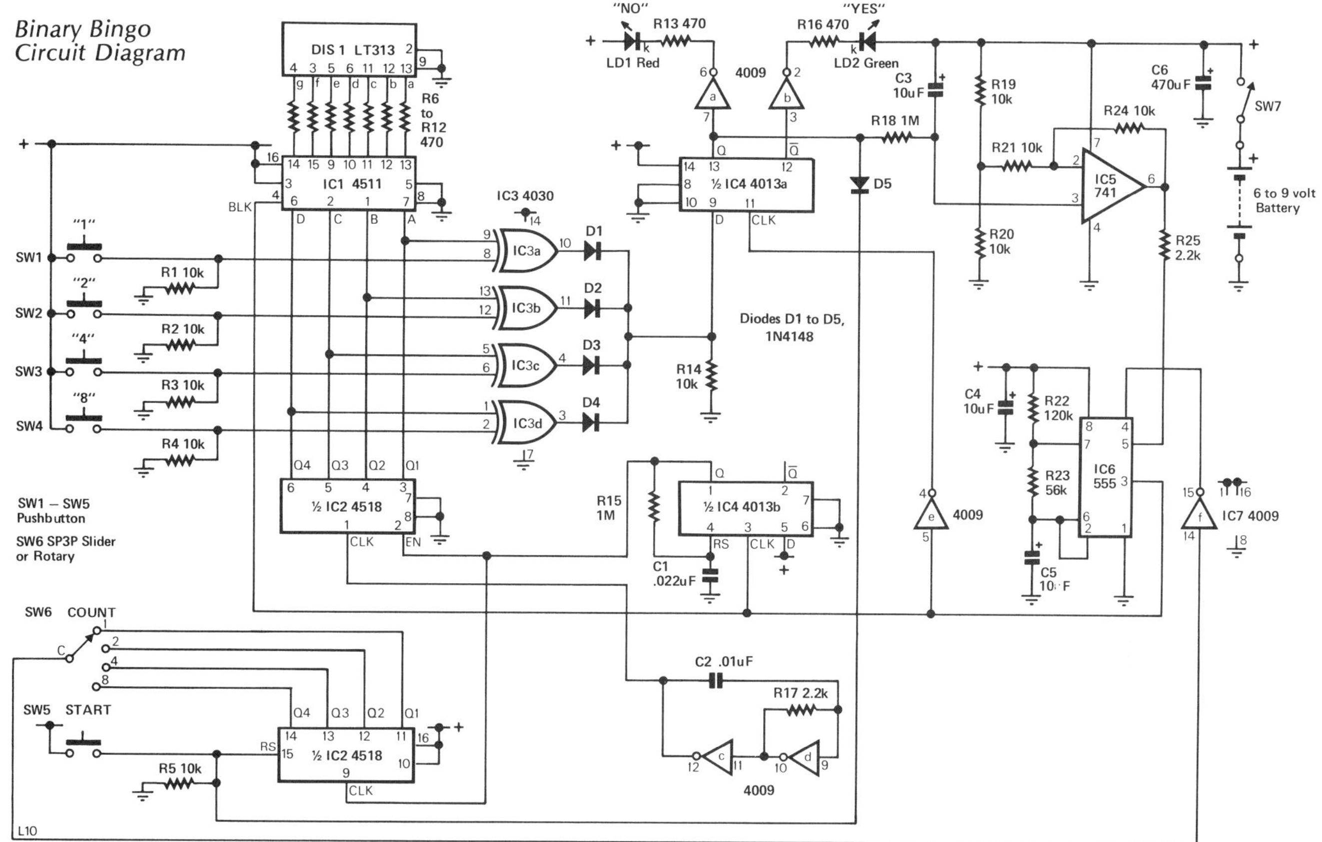

The circuit has two basic timing periods. The first and the longest is the decision period. During this cycle, a decimal number is displayed by the seven segment readout. At the same time, the binary equivalent of this number has to be keyed in and held by the four pushbuttons. During the second period, the display is blanked, at the same time the number keyed and the number appearing on the output of the decade counter are compared. If the comparision is the same, the 'yes' LED will light. Conversely the 'no' LED will come on if the two numbers are not the same. After this comparision is made, the decade counter is clocked by a free running oscillator. The system returns to the first state to repeat the cycle.

Decade Counter Clock And Display

The decade counter (half of the 4518 CMOS pack), a free running clock made from two inverters (4009 CMOS buffers), the seven segment LED display (LT313) and a variable gating system (4013 and 555) make up what may be loosely considered a random number generator.

During a short gating period of around 20mS from the output of a monostable (half of the IC4 4013 flip-flop configuration), the decade counter is clocked by the free running astable made from two inverters of the 4009 running at approximately 20kHz. At the end of this clock burst period, the last number in the count is stored in the counter and is present at its outputs Q1 to Q4. These outputs are connected to the 4511 decoder/driver for the display, at this stage blanked.

System Clock

A 555 timer is used to generate the two timing periods in the operating cycle. It is basically a low frequency oscillator with a variable mark-space ratio. The longest of these two periods, the decision cycle, is around 3 seconds and pin 3 is high. During this time, a decimal number is displayed and a binary equivalent has to be entered. For the second and shorter period, around 0.6 seconds, the output is low. This is the active system clocking period.

The decision period can be shortened by using the 555 voltage control pin 5. This is driven by a 741 op-amp that is used as a high input impedance buffer of the voltage present on the 10uF storage capacitor C3. This actual voltage can vary somewhere between the supply rails depending on the state of the decision memory flip-flop output pin 13 of IC4.

Let's assume that the last answer resulted in a true answer. The 'no' LED would be off and therefore the Q output, pin 13 would be low. This would result in C3 being discharged towards the negative rail via R18. This decreasing voltage is applied to the 555 via the op-amp buffer (current limited by R25). This decreases the threshold operating point (normally two thirds the supply rail) and therefore shortens the charging period of the timing capicitor C5, (10uF). This results in reducing the 555 on time and as a result shortens the decision period. From this action, we can see that if the correct answer is given, the decision time is reduced and the skill required to make these decisions is increased. Conversly, if a mistake is made, the 'yes' LED will be off and therefore pin 13 of the 4013 will be high. This results in an increase in voltage on C3 and as a consequence, the decision time will be increased. The range of this time period may range from around 3 seconds to less than 1 second. If a shorter period is required, this can be reduced by changing the 555 timing charge component R22. Try 100k down to 68k.

Correct Answer Counter

The second half of the dual 4518 is employed as a simple counter. It accumulates the sequential correct 'yes' answers and its outputs Q1 to Q4 are used to stop the system clock when a 'full card' is reached depending on the position of the 4 position switch. When the output connected via this switch goes high, it is inverted by the 4009 (f) and stops the 555 via the reset input, pin 4. This results in a 'winning card' and stops the game. The display is blanked and will remain in this state till the START button is pressed. This restores the counter to zero so that all outputs are low. The 555 is once again enabled. Any 'no' answer will reset this counter to zero via D5.

Comparator

During the decision period, the binary equivalent to the decimal number displayed on the readout has to be entered into the four pushbuttons SW1 to SW4. This number is compared to the binary output of the decade counter using four EXCLUSIVE-OR gates of the 4030. If the 'same' comparision is made, all outputs of the 4030 will be low and diodes D1 to D4 will be reverse biased. This results in the data input, pin 9 of the 4013a being held low by R14. If the switch combination was 'not the same' as the counter output, at least one of the 4030 outputs would be high and therefore pull the 'D' input high.

When the system clock changes to the active period, pin 3 of the 555 goes low. This signal is inverted by the 4009 inverter (e) and transferred to the clock input of the 4013, pin 11. It is on this positive going transition that the 4013 transfers the logic level on the data input, pin 9 to the 'Q' output, pin 13. If the 'D' had a low level from the comparator, a 'yes' answer would result on the green LED.

When the system clock reverts to the decision cycle, this positive going transition at pin 3 of the 555 will clock the 4013b mono via pin 3. As discussed previously, this short mono period will enable the decade counter clock system and at the same time, clock on the answer counter. The display will also be enabled as the blanking input of the 4511 goes high.

TO PLAY THE GAME

Switch the unit on and press the START button. Slide the COUNT switch to the 2 position. A number will appear on the display. Now press the appropriate switch or switches that would be the binary number equivalent. The table above these four pushbuttons show the equivalents if you are not familiar with the binary system. A '1' in the column indicates that the pushbutton has to be pressed on while '0' is off. eg; If the number on the display is 7, the '1', '2' and '4' switches have to be pushed and held. A 9 on the display would require the '8' and '1' buttons to be held on.

After the decision time period, the display will blank and one of the decision verification LEDs will light. If you pressed the correct binary equivalent pushbuttons, the 'YES' LED will light. Now the next decimal number will come up on the display; press the correct pushbuttons. If your answer is right, the 'YES' LED will remain on. The display will blank but will not come on again if the two answers were correct. This is because the answer counter has incremented two places, the number you first entered in the count switch. This indicates you have a 'full card' and 'Bingo', you have won the game.

Simple, you say! Now move the count switch to the 4 or 8 position, press the start button and play again. You will notice that as you get more answers right, the decision time becomes shorter and therefore gives you less time to think. It becomes much easier to make a mistake. Each time the answer is incorrect and the 'NO' LED comes on, the 'count' register is reset to zero. You can only win the game with a full card if all sequential answers are correct. Obviously, the 1 and 2 positions of this counter do not require much skill and are provided for beginners.

Where a mistake is made, the decision period is increased so you have time to concentrate. If you find that after a number of games you become familiar with the operation and become proficient, the decision time can be shortened. See the How It Works section for details of this circuit change.

This game is an excellent teaching aid to the binary number system while, at the same time, it tests and decreases your reaction time.

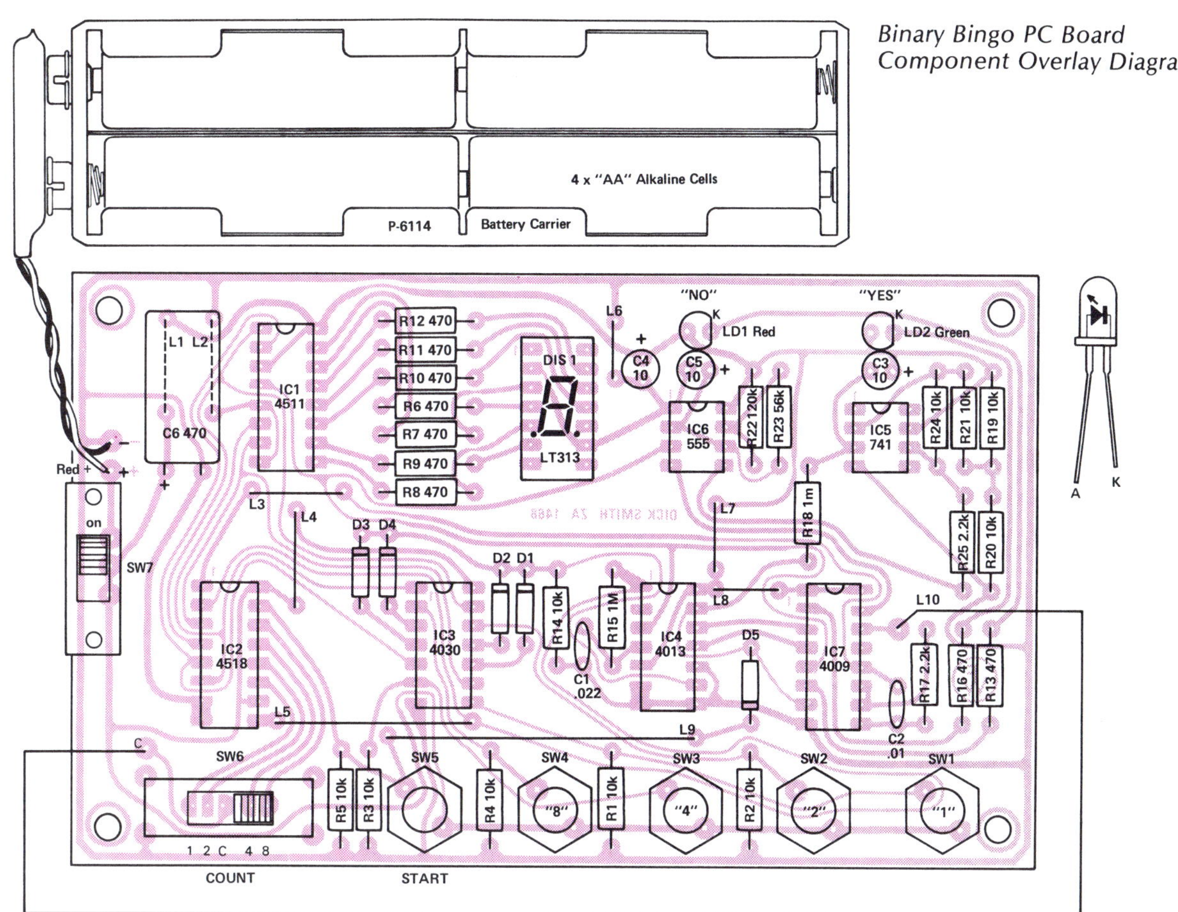

ASSEMBLY

1. Check off your kit of components against the parts list.

2. Load all low profile components. These include links L1 to L9. L10 is a length of hook-up wire. The resistors and diodes are also inserted. Make sure of the polarity of the diodes. Cut excess lengths and pigtails. Solder as necessary.

3. All capacitors can be inserted, pigtails cut and soldered. Note polarity of the electrolytics. The larger 470uF capacitor C6 can be laid on its side to reduce the height so that the board can be used in a Zippy Box.

4. The IC's can be carefully inserted and soldered. The battery lead with the snap attached can also be soldered. (Note red is positive.)

5. The final housing of the board has to be considered at this point. If you use the Zippy Box method we have suggested, the display can be inserted. Mount it centred in the 14 hole position. The device itself may only have 10 legs so that pins 1, 7, 8 and 14 are not present and these holes in the board are not used. Do not push the display hard up onto the board; only about 1mm of the legs should penetrate the foil side of the board. Solder in this position. This gives the display a little extra height so that it is closer to the window of the case. If you wish, an optional 14 pin IC socket can be used to increase this standoff distance. The display is plugged into this socket – again pins 1, 7, 8 and 14 may not be used.

6. Housing

An optional UB1 H-2751 Zippy Box can be used to house the board. The dress panel for the lid of the case can be cut from the rear of this book. The holes and cutouts for this panel can be marked by using the label as a template. Place it over the lid so that the edges line up with the marked outlines. With a sharp instrument like a scriber, carefully mark the centre positions of the 5 pushbuttons, the 2 on-off switch and the 2 LED holes. Mark each corner of the 3 rectangular cutouts and remove the label. Accurately scribe a line between each marked point of the cutouts so that a rectangle is clearly visible.

The holes can now be drilled at the centres marked, 6.5 to 7mm for the pushbuttons, 2 x 2mm for the on-off switch and 2 x 5mm for the LEDS. The rectangular sections can be cut out with a piercing or jewellers saw. This means that a pilot hole will first have to be drilled in the centre of the cutout to accommodate the saw blade. Do not cut right up to the scribed line; the final dressing of the

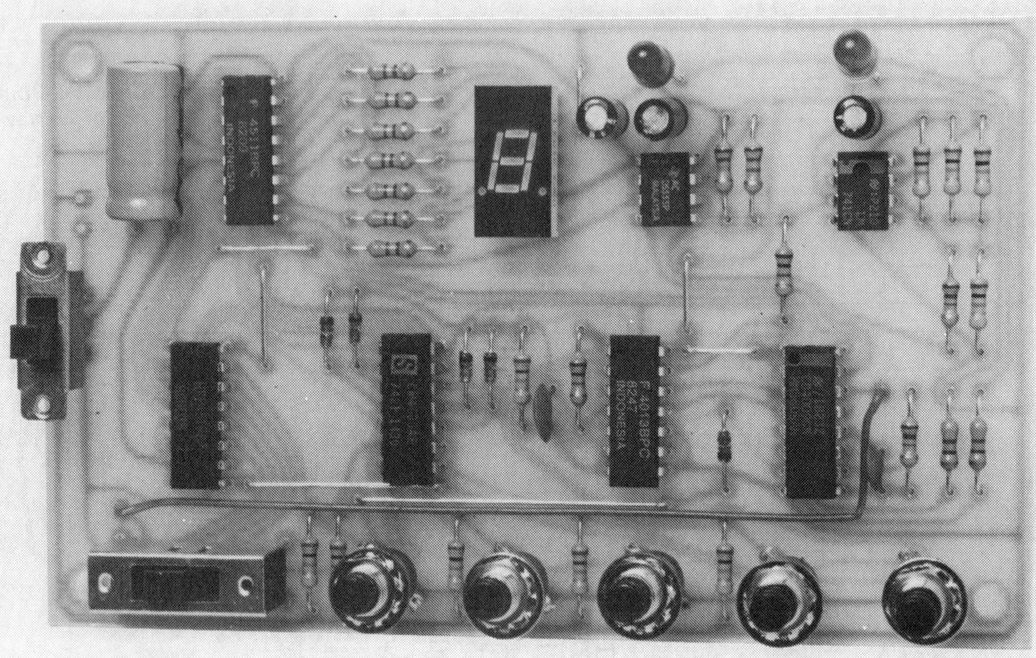

The complete PC Board showing the switches fitted.

hole can be done with a small needle file. If a saw of this kind is not available, a hole can be drilled in the centre of each cutout and then a small needle file used to enlarge and shape the hole to the final size. All burrs should be removed from each hole with a small file.

The label can now be glued to the lid by using a rubber based contact adhesive. Position it accurately before finally pressing down. The holes can be punched through this label by using the rear shank of the drill for the size hole. Push it straight through from the face side so that a clean cut is made. The rectangular cutouts can be trimmed by using a sharp razor blade or Stanley type knife. Trim around the edges of the label and lid if necessary.

The final PCB assembly can now be completed. The S-2060 count switch should drop neatly into the hole system in the board. It should be soldered with the body flush to the board. If the switch provided does not physically fit the same hole system, it may be necessary to wire the contacts and mount the body to the lid with screws. This means that two extra holes would have to be drilled.

The on-off and pushbutton switches can be mounted on the front panel with the screws and nuts provided. Do not tighten the nuts firmly at this stage. Solder short lengths of tinned copper wire to the contacts of each switch. Now insert the two LEDS in the appropriate holes. Note that the flat side of the body (k) faces away from the seven segment display. Do not solder at this stage.

The PCB can now be positioned over these switch wires so that they can be threaded through the holes in the board. If each pushbutton is turned till each contact is over the holes, the board will fit flush. The three wires from the on-off switch will have to be bent at right angles to accommodate the holes in the board. Depending on the pushbuttons supplied, it may be necessary to bend the end of the contacts so that the overall height of the switch is reduced. This then allows the lever of SW6 to penetrate the panel sufficiently. The nuts of the pushbuttons can now be tightened, not excessively but secure. The wires can now be soldered so that the PCB and the front panel are parallel. The LEDS can be pushed up to the front panel so that the shoulder of the

body is flush with the rear surface. Solder the four leads, and cut off all excess pigtail lengths. This completes all wiring.

7. The battery pack can now be plugged in and the unit tested. If all is well, the total assembly can be fitted into the Zippy box. If the unit fails to operate correctly, check all wiring, PC board assembly and soldering.

8. For those constructors who wish to make the assembly as solid as a rock, four standoffs can be used to fix the PCB to the bottom of the box. The length of these supports will have to be measured carefully so that the PCB, front panel and case have the right spacing between them. This extra work can only be justified if the unit is to be subjected to rough treatment. Normally, the simple wiring from the switches to the PCB is sufficient to support the system.

9. The P-6114 battery carrier and 4 x 'AA' cells fit between the bottom of the case and the PCB. A flat piece of foam plastic can be fitted between the PCB and this carrier to hold it in place. When the case screws are tightened, this foam should hold the carrier in position.

WHAT CAN YOU GAIN FROM BINARY BINGO

A knowledge of binary numbers is essential to understanding how and why computers do what they do.

You may not realise it, but every day you use binary numbers in a huge number of ways. And you may not even know what a binary number is!

This is because a huge variety of things these days have some form of computer built in. The trouble is, computers have a very limited vocabulary: they have only two words - off and on!

In a binary number system, there are only two numbers - a '0' and a '1'. Obviously, if the 'off' is called a '0' and the 'on' is called a '1', the computer circuitry can manipulate numbers.

Our binary bingo game teaches you binary numbers very quickly - and you have a lot of fun along the way! It displays a decimal number and you have to instantly work out the binary equivalent and push the binary buttons. If you get it right, it gives you a 'green' and gives you the next number - even faster!

A wrong answer gives you a 'red' and slows down the next number. When you have answered a set number times correctly, your 'bingo' card is full.

Sounds pretty easy, huh? Give it a go and see if you still think the same way!

Try to race you friends for a 'full card' or see how quickly you can do it yourself.

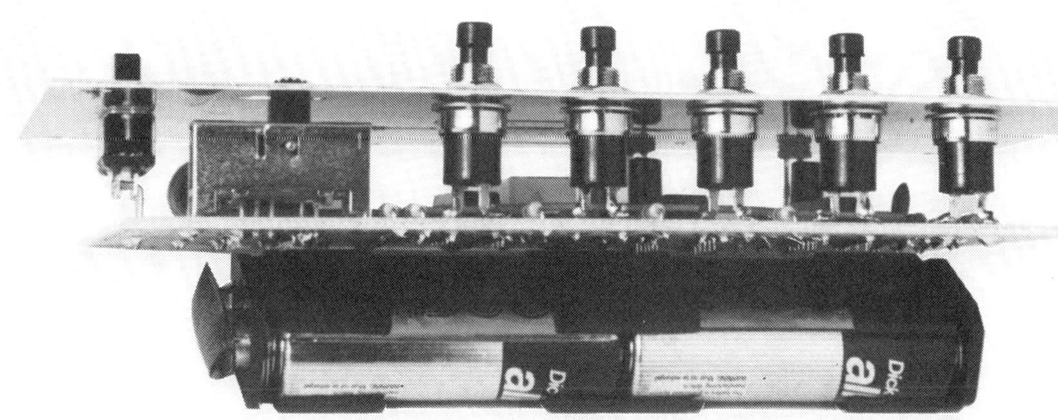

This photograph shows the complete assembly before it is inserted into the Zippy Box.

Mini Synth

This musical-based circuit is the most complex in the book, but it's a load of fun. By using a microphone sound detecting input, a whistle, hum or voice can be the note generator. In this way, you can be your own one-man-band (sorry, girls: one-person-band) without being able to read a musical note! Based on a Phase-locked-loop (PLL) design, the circuit 'reads' the incoming note frequency and 'holds' it until the next input signal. This note is available at the output of two channels determined by an adjustable decay circuit. Tremolo can be added as a further extension of the basic note which, incidently, can also be halved or doubled in frequency to give a multitude of different sounds.

A completed Mini Synth circuit shown in optional Zippy Box and components.

PARTS LIST

Parts Description		Total Quantity
Resistors:		
1k	R1, R19, R21, R36	4
4.7k	R2, R3, R8, R18, R26, R37	6
10k	R12, R14, R25, R31, R32	5
47k	R5, R13, R24	3
100k	R4, R6, R9, R10, R11, R15, R17, R22, R23, R27, R28, R33, R34, R39	14
470k	R20	1
1 Meg	R7, R16, R29, R30	4
4.7 Meg	R35, R36	2
10k	Trimpots VR2, VR3, VR6	3
100k	Trimpot VR4	1
1 Meg	Trimpots VR1, VR5	2
Capacitors:		
120pF	Ceramic C4, C8	2
.01uF	Ceramic C3, C10	2
0.1uF	Greencap Polyester C5, C12	2
0.47uF	Tag Tantalum C1, C7, C11	3

Parts Description		Total Quantity
2.2uF	RB Vertical Electrolytics C6, C9, C13, C14	4
100uF	RB Vertical Electrolytics C2	1
470uF	RB Vertical Electrolytic C15	1
Semiconductors:		
4013	CMOS IC5	1
4016	CMOS IC4	1
4046	CMOS IC2, IC3	2
LM324	Quad Op-Amp IC1	1
BC549/DS549 or 2N2222 NPN Transistors TR1, TR2		2
1N4148 Silicon Diodes D1 – D4		4
Miscellaneous:		
DP3T miniature Slider Switch SW1, SW2		2
DPDT Miniature Slider Switch SW3, SW4		2
Battery Snap		1

Parts Description	Total Quantity
Tinned Copper Wire 50mm	1
PC Board ZA-1469	1
Optional Components: (Not included in basic kit).	
10k Log Potentiometer VR2, VR3, VR6 – R-6820	3
100k Lin. Potentiometer VR4 – R-6810	1
1 Meg Lin. Potentiometer VR1, VR5 – R-6813	2
DPDT Miniature Slide Switch S-2010	1
Output Sockets eg. P-1231	2
Input Sockets eg. P-1231	1
Knobs, 25mm H-3860	6
Zippy Box Case UB1 H-2751	1
Tinned Copper Wire 22G eg. W-3022	1
Hookup Wire – assorted colours eg. W-4010	1
Basic Kit of Components. Cat K-2669	

OPERATION

This ingenious little unit samples and holds the frequency of a signal from a microphone or other sound source and outputs this signal as a single note. The duration and amplitude of the note can be controlled. The frequency of the note can be halved or doubled by the flick of a switch. A vibrato effect can also be applied by a variable tremolo control. To add to all of this, two channels are available to give chorus or stereo effect. Each of these channels has a separate volume control.

The signal source is derived from a low to medium impedance microphone. This microphone can be almost any dynamic or electret type. The types used with low cost cassette recorders will operate perfectly. The Dick Smith C-1015 Dynamic Type is a good example. Even a simple Electret Insert like the C-1160 will operate when correctly wired.

The synthesizer's output is connected to an external amplifier of some description. A home stereo system would be suitable with the auxillary or tuner input being the connection point. A PA amplifier could be used. The Mini Stereo Amplifier project in other sections of this book would be an ideal companion. Use the output stage section and keep the volume low at the start of your testing.

Connect the microphone and switch the unit on. Now slowly whistle a tune a short distance from the mouthpiece. Do not not blow directly into the insert. Best results are obtained with the microphone at the side of your mouth so that any air currents do not hit it directly.

Turn the input level control up till you get a reliable response from the unit every time a note is whistled at the same volume. Turn the output volume controls to a suitable level to avoid feedback into the mic. The whistled note should be a short clean burst. Try changing the decay controls from a short staccato to a long slowly decreasing tone. You will notice that the system outputs full volume as a noise burst is picked up by the microphone. You will also see that the synthesiser is sensitive to all sound. When the microphone picks up the output sound of the amplifier, feedback occurs and the system saturates into an uncontrolled state. You must keep the microphone away from the speaker system as with any PA system.

When you find a suitable input level, try the other functions. By changing the 'Octave Select' switches, you can select half or double, as well as the fundamental frequency. By adding tremolo to one or both channels, interesting tonal effects can be achieved.

Try singing or humming into the microphone as the sound source. The system will respond best to pure, clean notes. Rough or gravelly type voices will not be reliable and never make the 'Top 40'.

By striking different shaped objects next to the microphone, the system will tend to pick up the fundamental resonance of the object and produce an equivalent note.

Musical instruments as the sound source could be tried. A simple recorder as the signal source can produce many and varied sounds. Try a guitar: the combinations are endless.

To give greater versatility to the unit, channel one offers control from external sources. See the 'How It Work' section for details.

With a little practice, this Mini Synthesiser can make you a one man band.

The completed circuit board showing all switches fitted.

HOW IT WORKS

This simple synthesizer consists of four main sections. The input stage is basically a microphone amplifier and signal shaper with three stages. Twin voltage controlled oscillators (VCO) each with a voltage controlled amplifier (VCA) output stage makes up the main section. The remaining circuit block is a simple, adjustable low frequency oscillator to act as a tremolo effects generator.

Input Stage

The front end of this stage has been designed to accept the output from a low impedance microphone, either dynamic or electret. The quality of the microphone does not have to be high, in fact, the simple type normally used with a small cassette recorder will operate perfectly.

IC1a is the first stage having an AC amplification factor of around 100 times at 1kHz. The operational amplifier (op-amp) circuit is very basic. The only difference of any significance is the small capacitor in the feedback loop. This results in a roll-off in gain with an increase in frequency. The output of this amplifier is AC coupled to the next stage by C7.

Again, IC1b is a simple op-amp circuit with a voltage gain around 20 times. Like the previous stage, the high frequency response is limited. The response tailoring has been done to reduce the normally rich harmonic content of the human voice. The only signal we want to process is the fundamental note. The harmonics are low in amplitude and can be partially rejected by this simple approach.

The last section of the input stage IC1c is the most significant and needs some explanation. The op-amp is configured as a Schmitt trigger with offset. The hysteresis components are R22 and R21. Normally a Schmitt Trigger is bistable and in the quiescent state, the output would be either positive or negative depending on the last signal transition. The difference in this case is that the inverting input is offset greater than the value of the hysteresis voltage on the non-inverting input. Components R20 and R39 establish this offset to assure that the output at pin 8 is always low (negative) when no signal is present. The output will only switch when the input exceeds the sum of the hysteresis and offset

voltages. The circuit acts as a gating system to the following phase-locked loop circuits. This Schmitt trigger is the second stage in the processing to reject the harmonics and noise in the input signal from the microphone. The input level, set by VR2 is adjusted by turning this control until reliable response is achieved from a normal singing voice. It is not adjusted further than this minimum amount. In this way, the Schmitt will tend to only switch on the peaks of the fundamental. The harmonics and noise content of the signal are below the threshold switching points and will be rejected. The purer the voice or note, the more reliable the circuit action. A 'gravelly or rough' voice will tend to prove unreliable as a signal source. The best results will be from a whistle as this note is close to a sine wave and is relatively free from harmonics of any amplitude.

Voltage Controlled Oscillators

IC2 and IC3 are 4046 CMOS phase locked loops (PLL) and the main sections of these devices are used as voltage controlled oscillators (VCO).

Without input signal, the output of the Schmitt will be low and therefore pins 13 of IC4a (12 of IC4c) will also be low. This means that the 4016 analog switch will be off and pin 13 (PLL comparator 2) of the 4046 will be disconnected from the input to the VCO, pin 9. The voltage present on the low pass filter capacitor C5 (C12) will determine the frequency of the VCO in association with the RC timing components C3 (C10) and R5 (R24). This voltage will hold for a reasonably long period of time because of the very high input impedance of the CMOS input at pin 9 of the 4046 and the low leakage characteristic of the 0.1uF polyester capacitor. This in turn means that the output frequency of the VCO at pin 4 will 'hold' in this condition.

Now let us look what happens when an input signal occurs. Pins 13 (12) of the analog switch and 12 (signal input) of the PLL will go high with the signal. The phase comparator output (pin 13) of the PLL will be connected to the VCO via the low pass filter. Provided that the input signal continues at a fixed frequency for a short period of time, the PLL will lock and follow. In this locked state, the output of the phase comparator at pin 1 will, for the

greater part be high (except for narrow negative going pulses). For full explanation of the operation of these 4046 PLL systems, refer to manufacturer's data.

Voltage Controlled Amplifier (VCA)

With pin 1 of the 4046 high, D3 (D4) will be forward biased via R8 (R26) and so charge the storage capacitor C6 (C13). This voltage is applied to the base of TR1 (TR2) acting as a voltage follower/buffer. This transistor is half of the VCA output stage, the other section being another 4016 analog switch IC4b (IC4d). The control gate, pin 5 (6) of this switch is connected directly to the output of the VCO pin 4 or indirectly via the 4013 flip-flops. The square wave output from the VCO opens and closes this analog switch. It can be seen that this action directly gates the voltage available to the output terminal via the two current limiting resistors and variable control.

When the input signal disappears, pin 1 of the PLL returns low. As described previously, the VCO remains running at a frequency determined by the voltage on the capacitor C5 (C10). Now that the storage capacitor C6 (C13) has been deprived of its source, the voltage across the terminals will be discharged via the transistor follower and the two resistors, R12 (R32) and VR1 (VR5). These two components control the 'decay' rate. The control set to maximum value (1M) will result in a long discharge time (or another way of looking at it, little decay). As the discharge is taking place, the continuous output from the VCO switches the analog gate IC1b (IC4d) on and off to 'sink' the decaying voltage on the emitter of TR1 (TR2) via the load resistor R18 (R37) at the VCO rate. This is the basis of the VCA.

Output Frequency

The frequency of the note reaching the output stage can be changed from that of the original. By using a flip-flop as a divide by two element, the output can be halved or doubled depending on where it is coupled into the circuit. With the octave select switch SW1 (SW2) in the centre position F, the output frequency will be as the input or fundamental. Switched to F/2 position, the 4013 wired as a clocked flip-flop divides the output from the

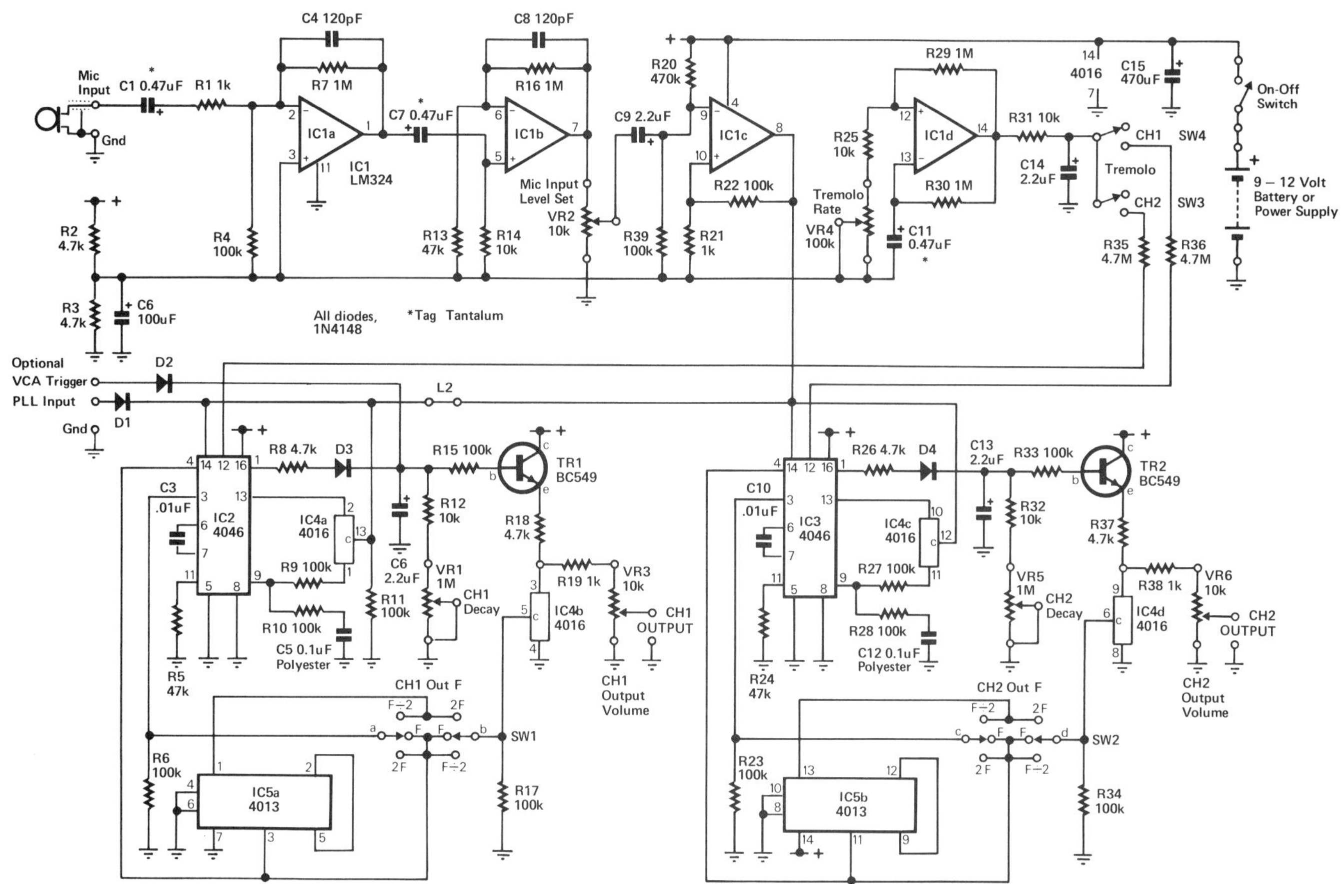

Mini Synth Circuit Diagram.

VCO pin 4 by two (lower octave). In the 2F position, the flip-flop is wired between the output of the VCO and the comparator of the PLL. This results in a frequency twice that of the fundamental (upper octave) at pin 4 of the VCO. (Again refer to the manufacturer's data on the 4046 for circuit operation).

Low Frequency Oscillator

The fourth op-amp of the LM324 pack is used as a low frequency oscillator to act as a tremolo generator. The circuit is configured as an astable multivibrator to give a square wave output. The variable control VR4 is included to adjust the frequency or tremolo rate. This waveform is then somewhat smoothed to give a more natural tremolo effect to the output note. This varying voltage is applied to the VCO via a selector switch and a 4.7M resistor. The VCO offset control input, pin 12 is used to slightly vary the fundamental frequency.

OPTIONAL INPUTS

VCA Trigger Input

The VCA of one channel can be triggered from an external source. A positive pulse input via D2 to the VCA will charge the storage capacitor C6. If this input is held high, the output from the VCA will also remain high as long as this voltage exists. If the input is a pulse from a sequencer or even a simple switch, the output from the VCA can be triggered without the need for an audio signal at the microphone input. This could be used with rhythm generators etc to create different effects. The frequency of the VCO can still be changed by using the microphone input. If you require that the input signal from the microphone amplifier stage does not trigger the VCA, remove D3.

PLL Input

By disconnecting the input stage from the VCO, this section can be used separately by providing an external input signal.

Isolation is performed by cutting the PC board track between the L2 pads. The input is applied via D1. This should be a square wave, no greater in peak amplitude than the DC rail voltage applied to the Mini Synth circuit and not less than 75% of this value (normal CMOS input switching characteristics).

This facility could be used to follow an external instrument or sound generator (provided the input meets the requirements of the CMOS 4046). By way of example, the output from a simple monophonic organ can be used to create more interesting tones and sound effects. The decay control will vary the note shape and the octave switch can change the note frequency.

To return the synthesizer to its original condition, the link L2 would have to be restored. For sake of convenience, a changeover switch could be used to enable this function to be operated externally.

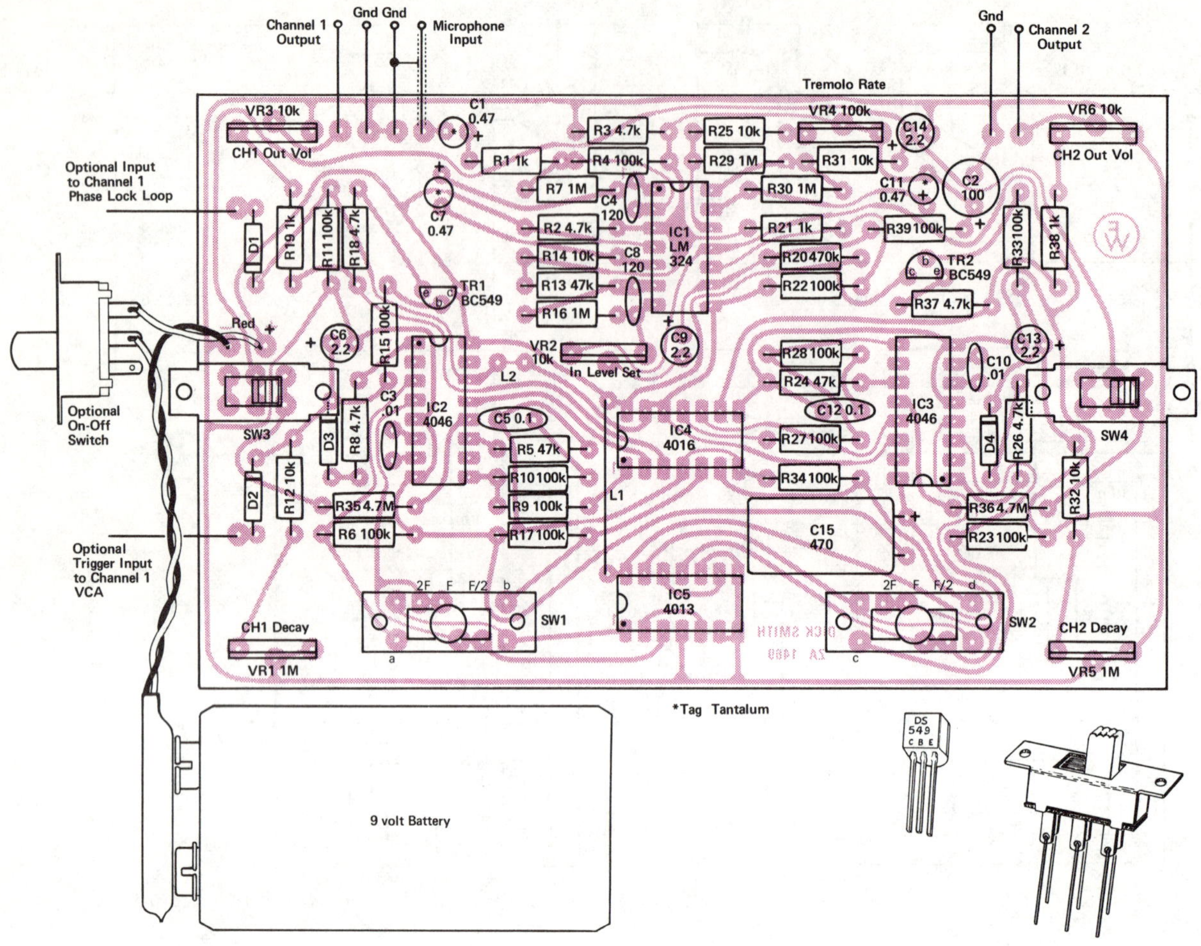

*Mini Synth PC Board
Component Overlay Diagram*

*Solder short lengths of copper wire to
each switch contact lug.*

ASSEMBLY

1. Check off your kit of components against the parts list.

2. Insert all low profile components into the board. Link L1, all resistors and diodes are loaded, pigtails cut and soldered.

3. All capacitors can now be inserted, leads cut and soldered. Remember that the electrolytics are polarised.

4. Load the ICs and transistors taking care with orientation, in particular, IC3.

5. At this stage, you have to decide if you are going to test and use the board in the "naked" form or put it into a case. To use without a case, the preset pots can be inserted as shown in the main overlay diagram. The power wiring, optional on-off switch, input and output wiring can be done.

6. **Switch Wiring.**
The board is designed to accept the four function switches directly above the surface. In this way, a UB1 Zippy Box can be used as the case. If you wish, these switches can be "soft wired" to suit the layout of whatever housing system you wish to adopt.

To position as shown on the overlay diagram, each of the contacts on the switches require short copper wires soldered to them. This enables them to be dropped directly into the hole system provided. When inserted, the contact lugs of the switch are flush with the top surface of the board. Tack solder one or two of the leads so that the switch is held in position. It now can be straightened so that it sits squarely and directly over the hole system. Now solder all leads and cut off excess. Each switch is now firmly located in position.

7. If you wish, the system could be tested at this stage. You will need an audio amplifier of some description as the output stage. The output level from this synthesizer is high so a high gain amplifier is not required. The output stages of the Mini Stereo Amplifier Project described in another part of this book is an ideal companion to this setup. If you wish, for testing purposes (and without disturbing the neighbours) a pair of high sensitivity headphones can be used to monitor the output. The type used with mini portable stereo FM receivers will work perfectly with the volume controls turned up fully. Use a suitable socket to connect if necessary. You may also need a socket to suit the microphone plug as well.

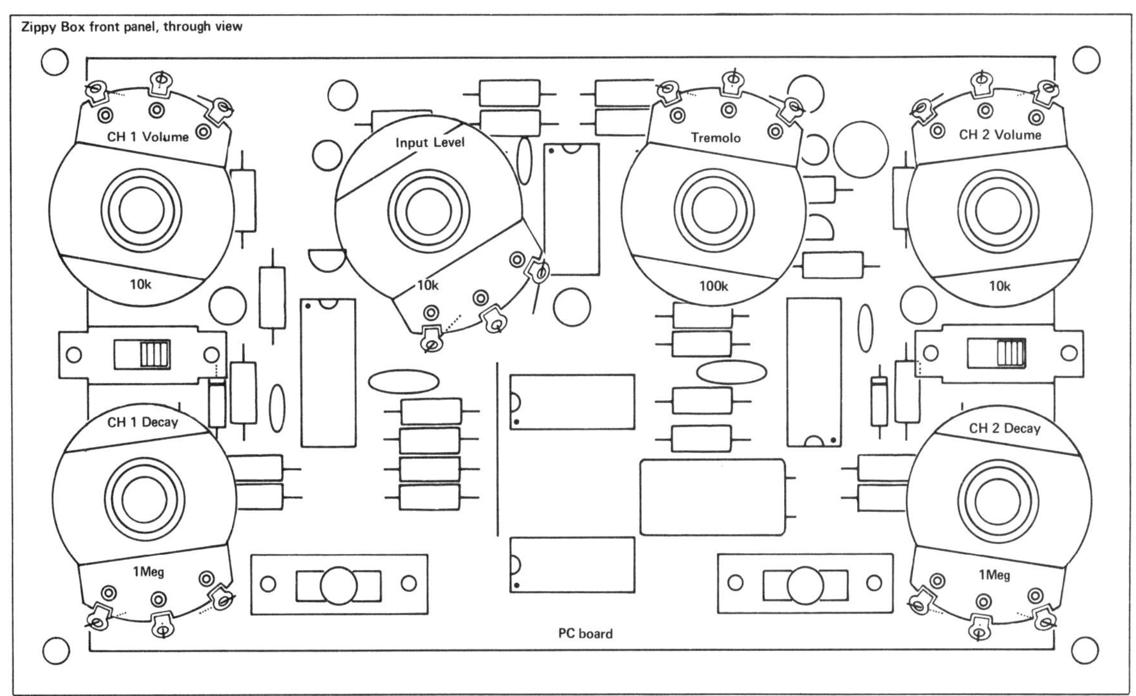

The complete circuit board showing the relative positions of the optional pots, when fitted to the front panel.

With all connections made, plug in the battery and switch on. Give a few whistle bursts close to, but not directly into the microphone. Turn up the Input Level Set control until some response is heard from the outputs. Turn Volume controls of both channels to suit level required. The output frequency should follow the whistle. Try changing the Decay controls from instant to very slow. With the controls set to maximum, you will hear the PLLs change in frequency as the note of your whistle changes. Notes that are wide apart in frequency will take a little longer to lock. Now try changing the Octave Select switches. Try the Tremolo Rate control and switches.

If all is working, the assembly can be completed. If something does not work, check all wiring, assembly and soldering on the PC board.

8. Case Assembly.

At the rear of this book is a suitable front panel label to suit this project and a UB1 H-2751 Zippy Box. This case system will require the components listed in the optional parts list.

Cut the label from the book and position it accurately over the lid. With a sharp pointed instrument such as a scriber, mark the centres of the pot holes. Lightly mark the corners of the switch cutouts and remove the label. Scribe lines between these points marked on the switch cutout positions so that the rectangular perimeter is formed.

The holes can now be drilled for the pots, 9mm to 9.5mm depending on the size of the control unit supplied. Drill a hole or series of holes in the centre of the switch cutouts to allow access for a piercing or jewellers saw. Cut inside the scribed line, dress with a needle file. If a suitable saw is not available, a needle file can be used to remove the waste.

Remove all burrs from the holes. The label can now be glued to the lid with a suitable rubber based contact adhesive. Position it accurately before pressing down. The holes can now be punched out by using the rear of the drill shank pushed through from the front surface. Use a sharp razor blade or Stanley type knife to cut out the switch holes.

Potentiometers will now have to be wired to the PCB to complete the assembly. If you had previously fitted trimpots to the board for testing, these will now have to be desoldered and removed. The overlay diagram shows the relative position of the pots with respect to the connections on the board. Short lengths of copper wire will have to be soldered to the lugs of each pot to make each connection. After soldering these wires, the pot can be loaded into the board at the position shown. Make sure you put the right value in the right position. You will have to bend some of the wires to suit the holes. Do not solder in these wires at this stage, simply bend over the excess to hold the pot roughly in position. When all six controls have been positioned, fit the front panel. This will take a little fiddling to align each pot through the hole provided. Fit a washer and nut to each shaft but do not tighten securely at this stage. Now move the circuit board around until it is parallel with the front panel and in the centre. All the wires leading to the pots should be free from other components and not touching the leads. Check that the two Octave Select and Tremolo switches move freely in the slots (move the board if necessary). The components on the board should have sufficient clearance from the pots to avoid touching. If all is well, the wires from the pots can be soldered and the nuts can now be tightened.

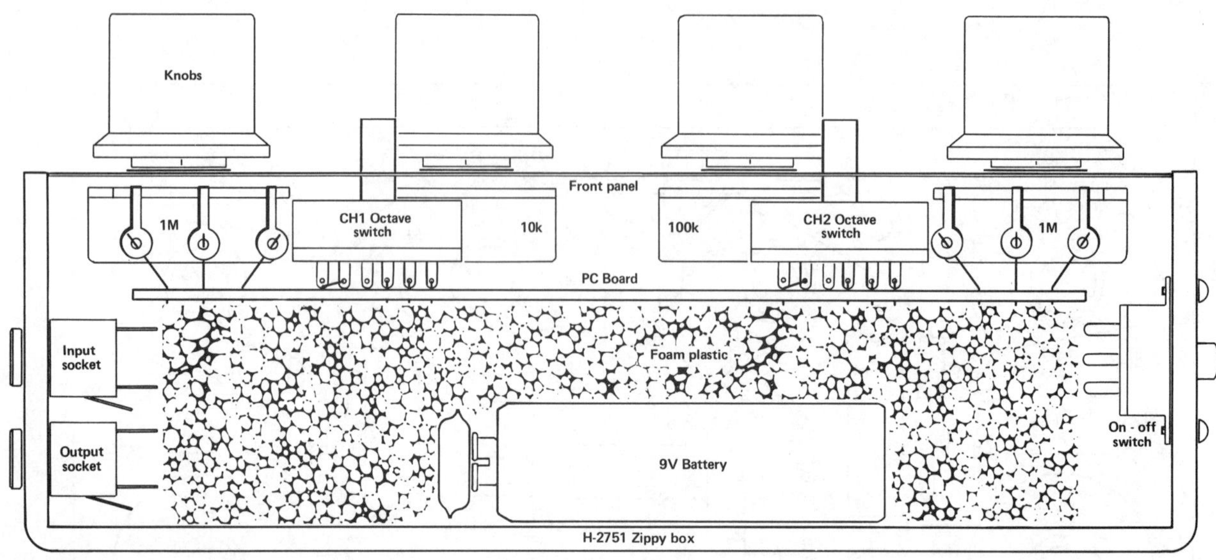

Inside the optional Zippy Box showing relative positions of all components.

CONNECTION OPTIONS

No provision has been made for an on–off switch on the front panel. Along with the input and output sockets, these could be mounted on one side of the Zippy box. The type of sockets used is up to the constructor. If the microphone used has a 3.5mm plug fitted, then it would make sense to fit the appropriate matching socket. The output connections could be RCA type as used on hi-fi gear or again the same 3.5mm type used on the mic socket.

As the synthesizer requires only a small amount of current (4mA to 7mA), a 9 volt transistor battery could be used as the power source. This could be located in the bottom of the box and held in position with a piece of foam plastic between it and the PC board.

When all connections are made and wiring is checked, the lid can be secured with the case screws.

Inside the optional Zippy Box, note the compactness of the design and how the pots neatly fit into and around the other components.

Technical terms

Aerial (antenna)
These terms are often used to mean the same thing. For our purposes, an aerial is a length of wire which is designed to receive or transmit radio waves.

AC
Stands for alternating current. Current which goes through a number of 'cycles' of reversal each second.

AF
Stands for Audio Frequency. It is generally accepted that audio frequencies occupy the part of the spectrum below 20kHz. Most humans can hear down to about 30Hz and up to about 15kHz but some animals, notably dogs and cats can hear higher than this. **(See Dog and Cat Communicator Project in Fun Way Vol. I).**

AM
Stands for Amplitude Modulation. One of the types of radio transmission (the type used by most broadcasting stations) where audio frequency signals vary the strength of a radio frequency 'carrier wave'. This is called 'modulation'.

Amplify
To enlarge. A circuit that amplifies is one that takes the original signal and **adds energy to it** without substantially changing the nature of the signal. Changing the nature of the signal is called 'distortion'.

Audio Transformer
A device which transfers audio signals from one section of a circuit to another, while blocking DC. **See 'Components' section.**

Base
One of the connecting leads of a transistor. **See 'Components' section.**

Battery
A device for supplying electric power. **See 'Components' section.**

Bipolar
A type of transistor, historically the first type invented. Bipolar literally means 'having two poles'. In a bipolar transistor the current through the emitter flows to or from (depending on type) two terminals – or poles – the base and the collector. **See 'Transistor' in the 'Components' section.**

Broadcast Band
The section of the radio frequency spectrum covering 'medium wave' radio stations – normally 530 to 1600kHz.

Capacitance
Having the properties of a capacitor. **See 'Components' section.**

Carrier Wave
A radio wave which does not have any signal information of its own, but carries the signal along with it.

Cell
A single battery or chemical power source. The batteries we use in these projects contain various numbers of cells to produce different voltages.

Circuit
The components, connections, wires and hardware that together form a working module. A drawing of the circuit is called a circuit diagram or schematic diagram.

Collector
Another of the leads of a transistor. **See 'Components' section.**

Conductor
An object which will allow electricity to flow through it relatively easily. Most metals are conductors; most non-metals are not. Some conductors are better than others.

Connection
Where two (or more) component leads or hook-up wires actually and deliberately touch, so that current can flow between them. If a connection occurs where it is not wanted it is sometimes called a 'short circuit' (or 'short').

Crossover
Where two component leads or hook-up wires intersect, but no connection occurs because they are insulated from one another.

Current
A movement of electrons along a wire or other conductor. By convention, current flows from the positive terminal, through the circuit and back to the negative terminal of the battery.

Dark State
When referring to a light sensitive device such as an LDR, the dark state refers to the behaviour or operating conditions when no light is falling on it. The converse is the device's 'light state'.

DC
Stands for Direct Current – current which does not change in direction.

Distortion
See 'Amplify'.

Earth (Ground)
Normally taken to mean the same thing. Can mean a direct connection to the earth itself, but is also used to mean a connection to a chassis or point of zero voltage.

Earphone
A device for turning electrical energy into sound waves. Called an earphone because it fits in or on your ear. **See 'Components' section.**

Electric Current
See 'Current'.

Electrolytic
When used in conjunction with capacitors, refers to the type of capacitors made with an 'electrolyte' **(See 'Components' section).** Electrolytes are chemical liquids or pastes having certain properties, used in various ways in electronics.

Emitter
Another lead of a transistor. **See 'Components' section.**

Feedback
Occurs when some or all of the output of a device (an amplifier, for example) can be fed back into the input. Feedback may be accidental and unwanted (as in the case of acoustic feedback from a speaker to a microphone with a resultant squeal) or deliberate (as in many types of circuits where feedback assists in correct operation).

Ferrite Rod Aerial
A coil or coils of wire wound around a rod made of ferrite (a black or grey material). Ferrite concentrates the effects of the coil.

Frequency Response This is the band of frequencies in which a component will efficiently operate.
Frequency response is a term commonly used in relation to HI-FI. A quality loudspeaker system, for example, in a suitable cabinet may claim to have a frequency response from 50Hz to 20,000Hz. To qualify this frequency response you must state several other factors e.g. you. must state whether the same amount of sound is radiating from the cabinet at 20,000Hz compared to say 1,000Hz or 50Hz. If one frequency is being reproduced more efficiently (i.e. "louder") than the others it must be specified by **how much** compared to a reference level. You must also state what power is being fed into the system. The "smoother" the frequency response the more likely it is that all frequencies within the band will be reproduced in accordance with the original signal.

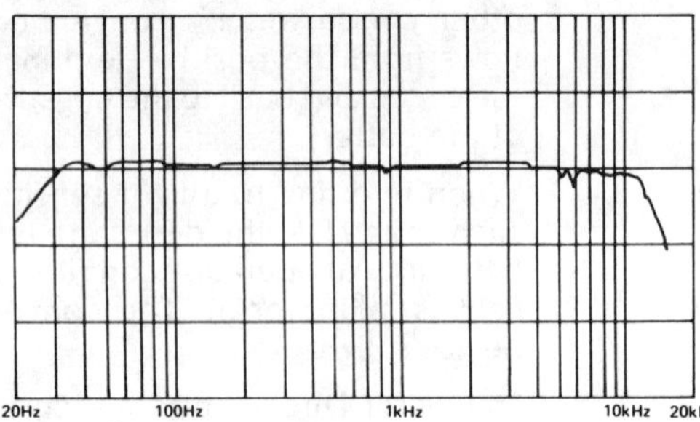

A close to ideal response graph; not really possible in practical terms.

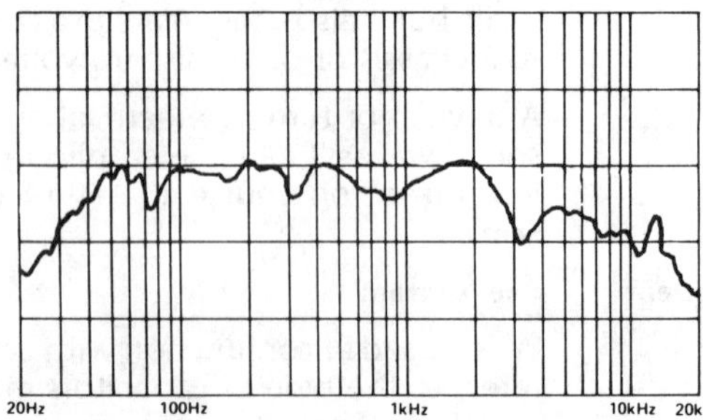

This shows the response of a speaker with peaks and troughs indicating that the sound would not be ideal.

Frequency The number of times an AC signal repeats its cycle each second. It is measured in Hertz (Hz). Was formerly called cycles per second. In other words, one Hz equals one cycle per second.

FM Stands for Frequency Modulation. It is another way of transmitting signal information on a carrier wave, with several advantages over Amplitude Modulation.

FM Band The section of the radio frequency spectrum covering radio stations which use 'Frequency Modulated' transmission. In Australia this band is 88 to 108MHz.

Gate A digital electronic circuit whose output state (high or low) depends on the state(s) at the input. There are three basic types of gates: 'AND' gates, where all inputs must be high before the output goes high (if any input goes low the output also goes low); 'OR' gates, where any input going high will also make the output go high; and 'NOT' gates, which simply invert the input state – if the input goes high, the output goes low, and vice versa.

Generator A device for mechanically or electronically producing electric current. Normally refers to a machine producing DC – but can also be used in other contexts.

Germanium A rare metalloid chemical element, from which germanium small signal diodes and certain transistors are made.

Greencap A colloquial name given to a type of polyester capacitor. **See 'components' section.**

HI-FI Stands for High Fidelity. Fidelity means faithful or accurate in detail. High fidelity is commonly applied to give a broad impression of the performance of music and sound reproduction equipment.
Unfortunately the term is used far too loosely and it suffers in the interpretation. What one person regards as HI-FI, another doesn't!

Hertz **See frequency.**

Impedance The total opposition (resistance) a circuit (or component in the circuit) offers to the flow of alternating current at a given frequency. If the frequency of the AC changes the impedance changes, up or down. Impedance is measured in ohms. A resistor on the other hand does not change its ohmic value when the frequency changes.

Input The side of the circuit to which a signal is fed for processing. (Converse: output).

Insulator A material which will not allow electricity to flow through it. (Converse: conductor). Most wire is covered with a plastic insulator to prevent the wires shorting to any other conductors.

Integrated Circuit A tiny piece of semiconductor material containing up to thousands of transistors, diodes, resistors, and other components. **See 'components' section.**

Leaking (a) electrolyte escaping from a battery, capacitor, etc. Often very corrosive.
(b) the action whereby a charge stored in a capacitor or battery slowly diminishes. Occurs because no components can be made 100% perfect.

LED Short for Light Emitting Diode. **See 'components' section.**

LDR Short for Light Dependent Resistor. **See 'components' section.**

Loudspeaker Another device for converting electric currents into sound waves. **See 'components' section.**

Matching Transformer **See audio transformer.**

Microphone	A device for converting sound waves into electric currents. **See 'components' section.**
Modulation	**See AM and FM**
Monitor	To continually examine. We use LEDs to continually examine parts of the circuits – they glow if current flows.
Morse Code	A system of communication invented by Dr Samuel Morse many years ago. It relies on the sending of only two types of sound – a short one and a long one. All letters and numerals are represented by combinations of these sounds.
Multivibrator	A popular circuit in electronics where two transistors are made to alternately turn on and off.
NPN	One of the types of transistor we use in this book (converse: PNP). **See 'components' section.**
Oscillator	A circuit which produces alternating current. Can be at any frequency but normally used in relation to audio or radio frequencies.
Output	That part of the circuit where the processed signal is available for use. (Converse: input).
PNP	Another type of transistor we use. (Converse NPN). **See 'components' section.**
Polarised	A component, plug, etc, which must be inserted or connected a certain way for the circuit to operate. If reverse connected, damage may occur.
Polarity	The specific way a polarised component must be connected to the circuit for correct operation (**see ''Polarised'' above**).
Polyester	A type of capacitor. **See 'components' section.**
Power	Normally taken to mean the source of energy which makes a circuit work (e.g. when power is applied). Electric power in our case.
Primary	(a) the 'input' side of a transformer. (b) a non-rechargeable battery, such as a flashlight battery.
Probe	Usually connected to test equipment to enable connection to various parts of the circuit under test.
Radio Frequency	RF for short. The part of the spectrum with frequencies from around 20kHz and above.
Relay	An electro-mechanical device which switches a set or sets of contacts when sufficient current is passed through its coil. The coil becomes an electro-magnet, pulling in a plunger which moves the switch contacts. (**See components section and 'using relays'**).
Schematic	Another name for circuit diagram.
Secondary	(a) the 'output' side of a transformer. (b) A re-chargeable battery, such as a car battery. You can feed electric current back into the battery and charge it. You must not do this with a flashlight battery.
Short	Short circuit. **See 'connection'.**
Semiconductor	(a) Neither a conductor nor an insulator, but somewhere in between. Often with the properties of being able to change by various means. (b) the name given to a whole family of devices using semiconductors for their operation: transistors, diodes, integrated circuits, etc. **See 'components' section.**
'Small Signal'	A device which is not meant to handle large currents. Such devices are easily damaged if overloaded.
Spaghetti	A form of insulation. **See 'components' section.**
Switch	A device for connecting and disconnecting power or other components to a circuit or parts of a circuit. **See 'components' section.**
Time Constant	In our circuits, a measure of the time it takes for a capacitor to charge (or discharge), with a certain resistor.
Timer	Any circuit or device capable of causing an action after a given time delay. Most of our timer circuits use a special integrated circuit designed to do this, with just two external components to pre-set the time delayed required.
Tolerance	All components are made so they are within a certain percentage of their marked value – this is the tolerance. For example, a resistor marked 1k, 5% might in fact have a value anywhere from 950 ohms to 1050 ohms.
Transistor	A semiconductor amplifying or switching device. **See 'components' section.**
Transmitter	A device for emitting radio waves. These may be bursts of carrier wave (such as a Morse transmitter) or they may be modulated by audio frequencies (as in a voice transmitter).
Tuning Dial	A dial connected to a variable capacitor, often marked so the stations can be easily found. **See 'components' section.**
Variable Capacitor	A type of capacitor constructed so that it can be varied by moving a rotating shaft. Usually used with some form of tuning dial. **See 'components' section.**
Volume	The level of sound which comes from a loudspeaker or earphone. A volume control is usually a potentiometer, and is situated at the input to the main amplifying section of the circuit.

About the 'binary' system

You will have noticed in the Binary Bingo that we talk about the binary system. Do you know what this means and how it is used?

In the number system we use in everyday life, we use ten symbols: **0, 1, 2, 3, 4, 5, 6, 7, 8, 9** to represent a certain numerical quantity. If the quantity is greater than any of these symbols will allow, we combine them to form a group of symbols, with the position of each symbol within the group having a meaning, as well as the symbol itself. Because there are ten symbols, this system is called a 'decimal' system, or a system to base '10'.

The reason we use a system based on ten is probably because we have ten fingers: but it is quite possible to count using other systems. Indeed, over the centuries various civilisations have used number systems based on 5, 12, 15 and many other numbers.

So it should come as no surprise to find that it is possible to count in a number system based on two: this is called the **'binary system'**. The most obvious difference between a binary and decimal number representing the same quantity is that the binary number is much, much longer . . . and it has only two different digits in it!

There is a big advantage is having just two digits. These are very easy to represent electronically. By convention, when talking 'in binary', we call the first digit '0' and the second digit '1', but we could (and do) call them other things: 'off' and 'on'; 'high' and 'low'; 'open' and 'closed' for example. Electronic circuitry to represent just two states is very easy to make, and it can be made with high accuracy – much better than for a circuit needing to recognise ten states, as it would if it worked in pure decimals.

Recognising and working with binary numbers is just as simple as working with decimal numbers (if not simpler!) To see what we mean, let's examine the same number in both decimal and binary notation: decimal 1387 (which can also be written 1387_{10}, meaning 1387 to base 10), or binary 10101101011 (or 10101101011_2) – we said that binary numbers are a little longer than decimal numbers!

Refer to the tables at the bottom of the page. They show both these numbers – let's look at the decimal one first.

Each digit in a decimal number has a 'weight': depending on its position. Each digit to the left is one power of ten higher than its right-hand neighbour.

So the number 1387_{10} really means: 'multiply 7 times 1; 8 times 10; 3 times 100; 1 times 1000; then add them all together'.

Now the binary number. It means just the same thing: but is actually simpler because there is no multiplication needed: each digit represents a power of two, or zero.

So that huge number really means, quite simply, 'add together $1 + 2 + 8 + 32 + 64 + 256 + 1024$'.

See how it works? It really is very simple!

Converting one to the other

It's easy to convert a binary number to decimal: we've just done it! To start off, you'll find it much easier to write down the number in a table form, just as we have done below. Then list the powers of two starting from the right (don't forget 2^0 is first – and any number to the power of '0' equals 1).

Now, let's try converting a decimal number into binary. There are a couple of ways to do this, but there is a way for beginners which is very simple: all you need to do is know how to divide by two! It works like this:

Take the decimal number, and divide it by two. If there is a remainder (in other words if the decimal number was odd) write down a '1' on the right of your sheet of paper (so you can write more numbers down on its left). If there was **no** remainder, write down a zero.

Now keep on repeating the division by two of each answer you get. Once again, if there is a remainder, write down a 1, no remainder, write down a zero. Keep on writing the numbers leftwards across the page, until you can't divide any more. If there was a remainder left over (it would have to be 1!) write down a 1, if not forget about the last zero. That's your binary number!

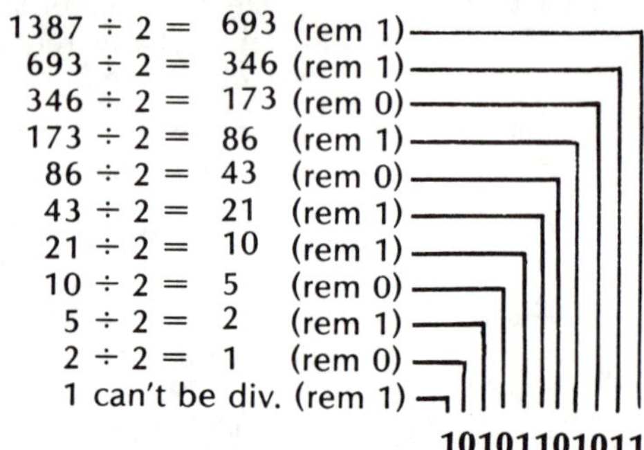

Addition and subtraction:

It's just the same as decimal arithmetic. You simply add the digits together, and if the number is greater than the base (in this case 2) you 'carry' the extra digit over to the next column. Subtraction, of course, is the reverse procedure.

Multiplication and division

While they are the same in theory, most digital circuits don't actually multiply and divide. To save memory and time, they use special circuitry: to multiply 50 by 20, a computer might add 50 to itself 19 more times (all in binary, of course!)

However, for the sake of the exercise, binary arithmetic is basically the same as digital arithmetic.

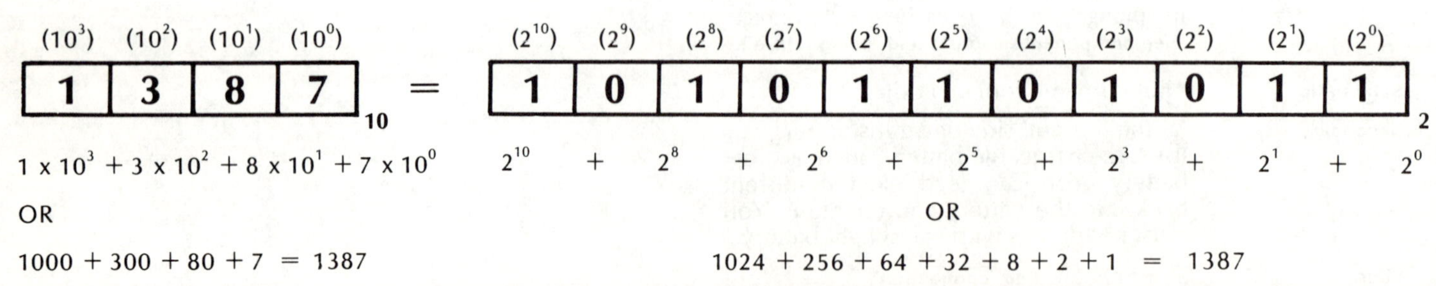

(10^3)	(10^2)	(10^1)	(10^0)
1	3	8	7

$= 1387_{10}$

$1 \times 10^3 + 3 \times 10^2 + 8 \times 10^1 + 7 \times 10^0$

OR

$1000 + 300 + 80 + 7 = 1387$

(2^{10})	(2^9)	(2^8)	(2^7)	(2^6)	(2^5)	(2^4)	(2^3)	(2^2)	(2^1)	(2^0)
1	0	1	0	1	1	0	1	0	1	1

$2^{10} + 2^8 + 2^6 + 2^5 + 2^3 + 2^1 + 2^0$

OR

$1024 + 256 + 64 + 32 + 8 + 2 + 1 = 1387$

Notes

Notes

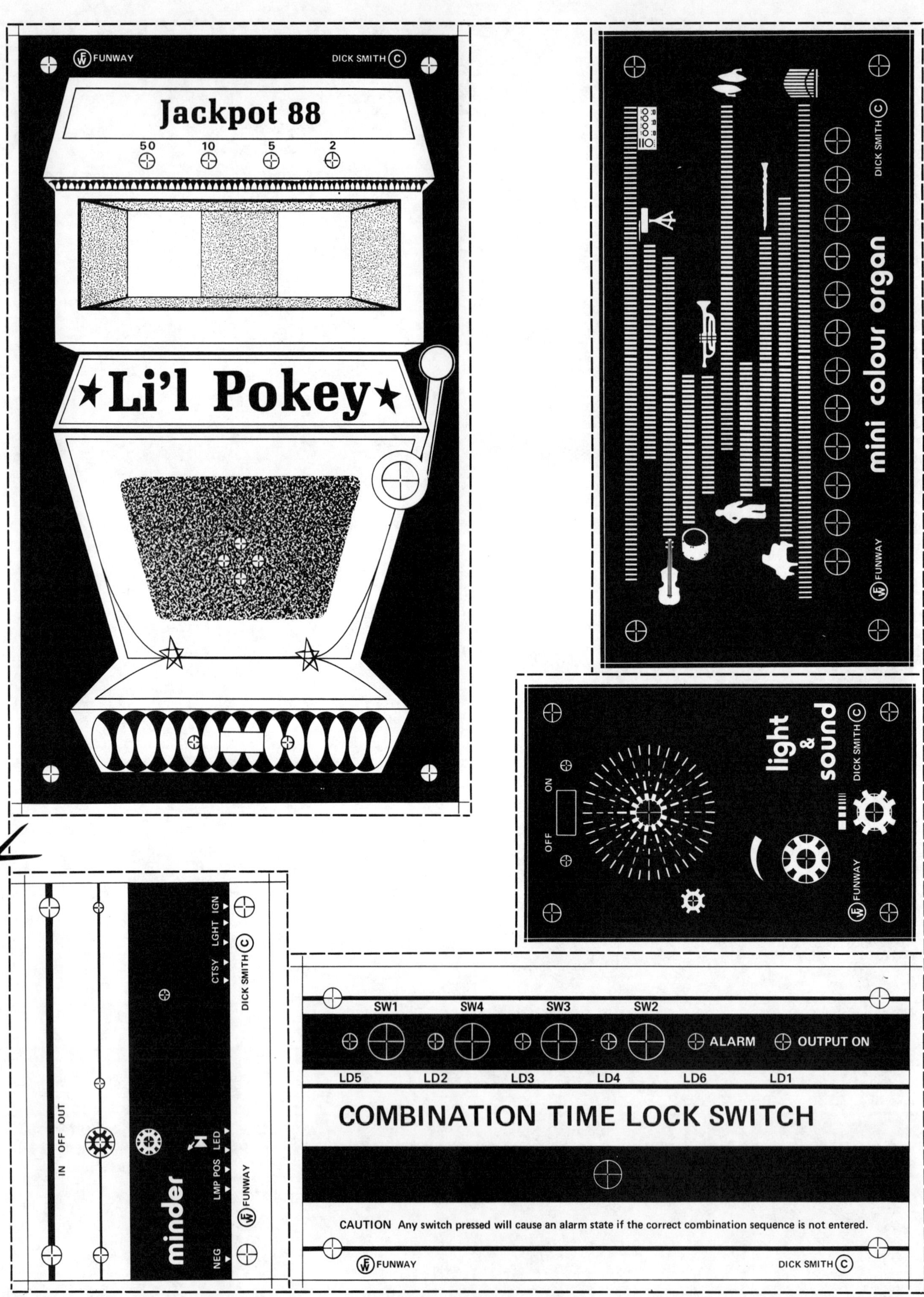
FUNWAY
DICK SMITH C
Jackpot 88
50   10   5   2
★Li'l Pokey★
DICK SMITH C
mini colour organ
FUNWAY
light
&
sound
DICK SMITH C
ON
OFF
FUNWAY
DICK SMITH C
CTSY LGHT IGN
IN OFF OUT
LMP POS LED
minder
NEG
FUNWAY
SW1   SW4   SW3   SW2
ALARM   OUTPUT ON
LD5   LD2   LD3   LD4   LD6   LD1
COMBINATION TIME LOCK SWITCH
CAUTION Any switch pressed will cause an alarm state if the correct combination sequence is not entered.
FUNWAY
DICK SMITH C

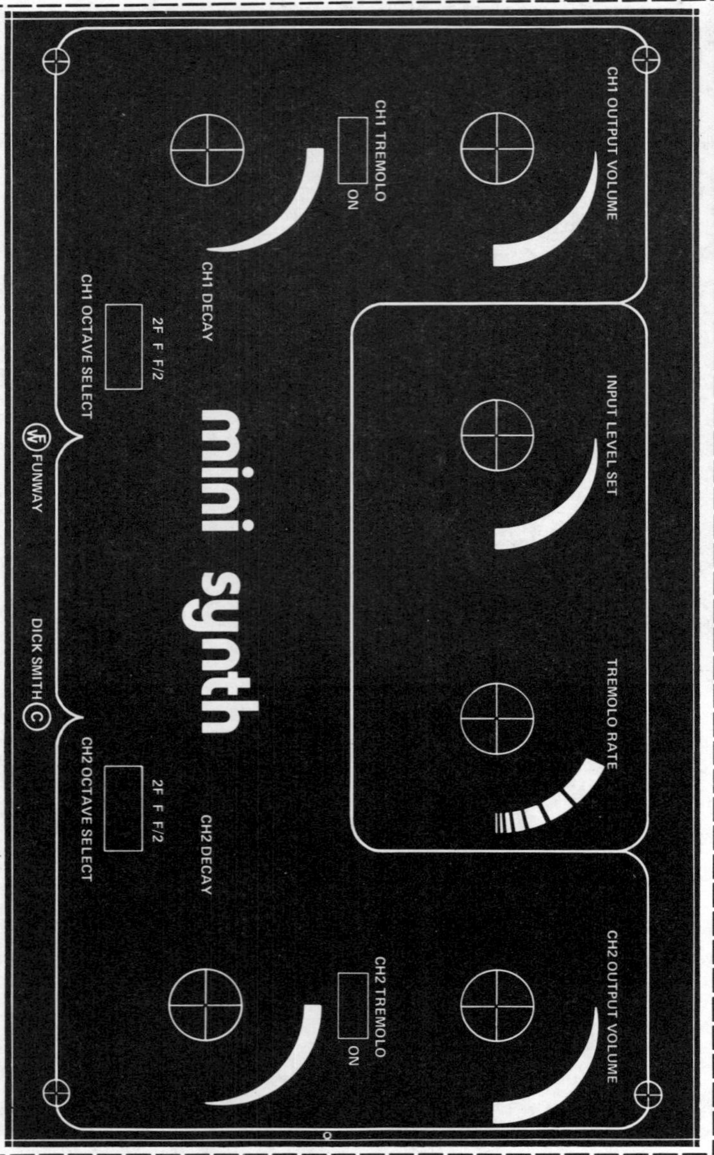

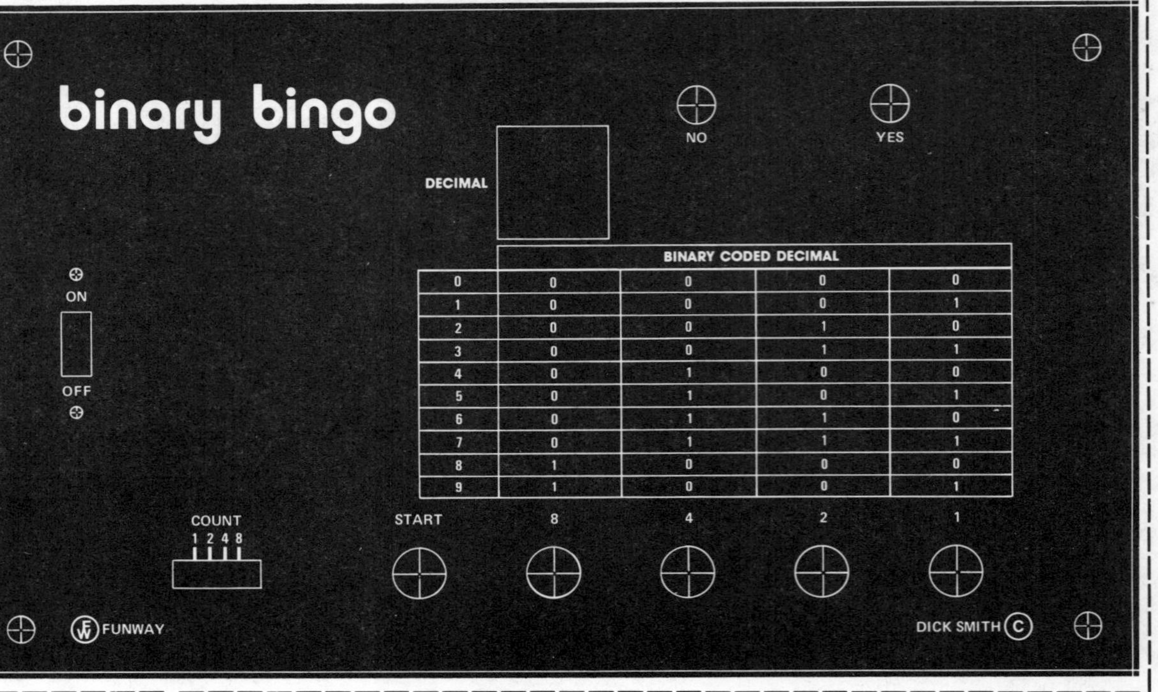

	BINARY CODED DECIMAL			
0	0	0	0	0
1	0	0	0	1
2	0	0	1	0
3	0	0	1	1
4	0	1	0	0
5	0	1	0	1
6	0	1	1	0
7	0	1	1	1
8	1	0	0	0
9	1	0	0	1

USE THE LABEL AS A TEMPLATE

Place the label accurately over the Zippy Box lid so that the dimension marks line up with the edges. Hold the panel and label up to the light so that you can see when both are in register. Now while still holding the two together, mark all hole centres with a sharp pointed instrument such as a scriber. Rectangular cutouts can be marked at each corner. Remove the label and scribe lines between points marked on the rectangular cutouts. Next drill or cutout each hole. Dress if necessary with a needle file.

TO CUT OUT THE LABEL

To remove the label required, cut along the dotted line near the perimeter. Do not cut right up to the thin dimension lines at this stage. These are used to align the Zippy Box Panel.

FINAL DRESSING

Coat the rear of the label with a thin film of rubber based contact adhesive. Place this label accurately over the Zippy Box lid and smooth down to remove all air bubbles. To give your panel a professional finish and protect the label, a layer of clear Contact or other similar self adhesive transparent film can be laid on the surface. The label can be trimmed to the final dimension by running a sharp knife blade around the perimeter of the panel. The holes can be punched out by using the shank of the drill of the same size as the hole. The rectangular sections can be removed with a sharp blade.